SYMBIOSIS

NICOLETTE
PERRY

SYMBIOSIS

Artwork by Paula Chasty

Close Encounters of the Natural Kind

BLANDFORD PRESS
POOLE · DORSET

First published in the U.K. 1983 by Blandford Press,
Link House, West Street, Poole, Dorset, BH15 1LL.

Copyright © 1983 Blandford Books Ltd.

Distributed in the United States by
Sterling Publishing Co., Inc.,
2 Park Avenue, New York, N.Y. 10016.

—15752—Blandford Press

British Library Cataloguing in Publication Data

Perry, Nicolette
　Symbiosis.
　1. Symbiosis
　I. Title
　591.5'24　　　　　　QH548

ISBN 0 7137 1229 5

Typeset by Asco Trade Typesetting Ltd., Hong Kong
Printed in Spain by Graficromo

CONTENTS

ACKNOWLEDGEMENTS

The author would like to thank Professor D.C. Smith of Oxford University for his help and advice.

The publishers are very grateful to Paula Chasty for the line drawings and colour artwork; and to the following for permission to reproduce colour photographs:
Bruce Coleman Limited (pp. 15, 18, 26, 30 upper, 31, 38, 51, 82, 94)
Ardea Photographics (pp. 19, 22–3, 43, 78–9, 83)
Biofotos (pp. 34, 54, 70–1, 87)
Oxford Scientific Films (pp. 2, 30 lower, 46–7, 75, 103)
Natural History Photographic Agency (pp. 39 upper and lower, 115)

INTRODUCTION

The interdependence of all living things on one another is well known. The world of relationships between organisms is finely balanced and anything that alters that balance prejudices life on a large or small scale. Natural and man-made events are constantly tinkering with the balances between organisms. These events can range from climatic changes (droughts and floods) to deliberate actions by man (cutting down hedgerows, clearing woodland and flooding valleys). After such events it can take decades before plant and animal life returns to normal. An event that is intended to affect one organism will inevitably affect many others. For example, if most insects are destroyed in a given area the local birds will have nothing to feed on and either starve or move away. Even in this simplistic example the cumulative effects for creatures that predate on the birds can be just as traumatic and damaging as those suffered by the insect.

Humans rely totally on plants and animals for life, but because they are omnivorous they are less vulnerable to the destruction of individual foods than those creatures which feed on only one specific plant or animal. The relationship between an animal and its food may be seen as a clearly one-sided affair, but there are many other examples where organisms co-exist with others for mutual benefit. This kind of relationship is surprisingly common and often very intense. These associations are frequently economic of effort for both parties; together they can survive much more efficiently than they could separately. The economy of these relationships can help otherwise vulnerable creatures to survive in a world where nothing is static and the natural balances are always being tipped one way or another. This beneficial association often makes the two partners together more viable than the sum of the two acting alone. There are cases of animals and plants that would surely have become extinct long ago were it not for the help of their partners; for example, blind shrimps being led around by sighted fish, flowering plants that need to be pollinated by specific insects, and even cows and other ruminants that cannot digest grasses without the aid of gut bacteria. These types of relationship are referred to as symbiotic, and are found not only in the animal kingdom but also among plants and even bacteria, algae and fungi.

As the word 'symbiosis' may not be familiar to every reader, an explanation of this term is needed. A mycologist named Anton deBary first coined the word symbiosis in 1876. The original definition was that a relationship was symbiotic if it was between dissimilar species, constant and intimate. He did not exclude relationships where one or more parties were actually harmed by the association. The inclusion of such parasitic relationships has since often been dropped from discussions on symbiosis, as it is usually dealt with separately. The number of relationships that are to a greater or lesser extent parasitic is so vast that they merit separate study. For the purpose of this book 'symbiosis' is taken to mean any mutually beneficial association between two or more dissimilar organisms. This includes fleeting relationships as well as long-lasting ones.

There are also other kinds of associations between organisms. They can associate together without causing each other any harm, but equally with no real benefit being gained by either party. Many skin micro-organisms fall into this category; they just live on the skin of different animals, neither aiding nor injuring the host in any way. As already mentioned, in other instances some species actually injure the other in various ways. This has come to be called parasitism, and there are many thousands of examples of this in both the plant and animal kingdoms. Mistletoe, familiar to us all at Christmas, is parasitic on various trees. It drains nutrients and water from its host and can cause the tree to slow or stop growing, and in extreme circumstances the host may die. In the animal world one of the most common parasites is the flea, which acts in a similar way to mistletoe. The flea punctures its host's skin and sucks out blood and therefore nutrients, causing the host to weaken.

This book is concerned with beneficial relationships between animals and plants of any kind. Some of these are at microscopic levels and others apply to much larger species, as these relationships span all types of flora and fauna.

Symbiosis is often a very convenient mode of life; if two species by living together can help each other feed, or protect one another, then both have a better chance of survival. This economy of effort is demonstrated over and over again in the different examples of symbiosis later described. Some species have become so reliant on another that they stop functioning as wholly separate entities. One species may totally rely on the other for its food supply, for example termites and their gut micro-organisms (*see* Chapter 7), or they may be unable to function properly in some way without the other. At the opposite end of the spectrum of symbiotic behaviour, some individuals meet only for a short time and help one another with one specific problem, for example the yucca moth which assists in the pollination of the yucca plant (*see* page 80). There are therefore many different degrees of symbiosis ranging from the casual meeting to total interdependence. Some associations are between two animals, some between two plants and some between animals and plants. Many of the relationships concern just two species but some involve several different organisms. Some associations are between two individuals and some are between whole populations.

It is therefore true that the range and depths of these associations is vast and covers every conceivable environment and condition. Symbiosis is thus of considerable ecological importance, and several of these relationships are important to the well-being of all living things.

The global importance of symbiosis cannot be over-stressed. All the animals that we refer to as herbivores (cattle, sheep, goats etc) depend for their own survival entirely on symbiotic organisms living within their bodies. Herbivores are those creatures that turn the raw products of grass, grain and other crops into meat and milk. These animals have been domesticated for thousands of years to enable man to survive through winters, and through both good and lean years. To enable these animals to provide man with meat and milk the animal has to be able to digest and convert its foodstuffs into usable protein and carbohydrates. Without this ability its food would be useless and the animal could neither grow nor multiply. Strangely, vertebrates in general are unable to break down cellulose, the basic element of grasses and many other plants, by their own digestive processes into these essential nutrients. Because of this inability to digest properly the grasses that they eat, all herbivores depend on intestinal microflora for their survival. These microflora consist of a dense mixture of many bacteria and protozoa which are only ever found in the digestive tract. These tiny organisms are able to break down the cellulose and thereby provide their host with essential nutrients in exchange for a constant and reliable habitat. Therefore it is on these tiny organisms that all meat, milk and cheese eating humans indirectly rely for the staple foods of their diet. The subject of herbivore digestion is discussed more fully in Chapter 7.

Other symbiotic relationships are of great value to man and all living things. We know that oxygen levels surrounding the earth are of fundamental importance to all creatures living on its surface. Without constant and adequate supplies of oxygen, all living things would suffer or die. Oxygen is a by-product of the photosynthesis process carried out by green plants, and therefore huge areas of forest and grasslands are required to supply enough oxygen for the whole world. Most of this oxygen comes from vast expanses of forest in, for example, northern South America and Canada. In a forest trees grow closely together, and as the roots below a tree use up at least as much space as the branches above it is not difficult to imagine the struggle for root space below ground. If, because of this proximity, the roots are restricted, the tree receives a reduced nutrient and water supply from the ground, causing it to slow its growth and in extreme circumstances to starve and die.

A symbiotic partnership comes into play in these circumstances between the tree and various fungi. The fungi weave 'webs' of mycelium around (and sometimes penetrate inside) the root tips of the trees. These webs, or 'nets' as they are sometimes called, effectively increase the root area of the trees, as the fungi then act as roots themselves. They take nutrients and water from the soil in exactly the same way as normal roots. These webs are very fine and able to spread into the most cramped soil spaces.

The action of the fungi means the difference between life and death in some dense forests, and even in sparsely populated forests they greatly assist the tree hosts to survive and grow. Therefore the oxygen supplied by dense tropical forests is largely dependent on the symbiosis between the trees and these fungi. The topic is discussed more fully in Chapter 8.

It might be worth pointing out at this stage that the thought of a reliance on fungi and intestinal microflora being in part responsible for all the meat and oxygen available to mankind would have been unthinkable a few decades ago.

Symbiotic organisms of various types also assist in turning leaf litter into humus which nourishes the original plants or trees that shed the leaves. This is vital to the well-being of plants and indirectly to all living things. Imagine a world in which lifeless vegetation did not rot away but built up and up. It would not take too many years to swamp all the plants and animals on the earth. The same is of course true of dead animal matter; various organisms both large and small deal with animal corpses, and thus remove these highly dangerous sources of pollution.

These examples of symbiosis are obviously of extreme importance to mankind. They are proven and easily demonstrable instances of the importance of symbiosis. The next example, in which it is suggested that the higher forms of life were only evolved as a product of a microscopic symbiotic relationship, is also of fundamental significance, although it is at the moment only hypothesis and, however well argued, remains speculation.

Nearly all organisms on our earth either have cells containing nuclei or cells with no nuclei; the former are referred to as eukaryotic and the latter prokaryotic. (An exception to this are the viruses, but as they can only reproduce inside other organisms' cells they are not relevant to the argument.) There are considerable differences between these two types of cell, as well as the nucleus already mentioned. The eukaryotic cells have a number of organelles, such as mitochondria and chloroplasts, which do not occur in prokaryotic cells.

Evidence on the relative ages of these cells has been found in the fossil record. Prokaryotic organisms long pre-date eukaryotic ones and it would also seem that eukaryotic cells appeared very suddenly. This would indicate that the commonly accepted theory of the process of evolution (step by step) did not happen in this case, but that the eukaryotic cells had some quicker means of production. This is backed up by the lack of modern 'intermediate' organisms in the form of creatures with 'primitive' organelles.

Most of the animals and plants that we are familiar with belong to the eukaryotes, each of their cells having the familiar nucleus in which the genetic material is contained. The prokaryotes exist in vast numbers but because of their relative size they are less familiar to us. They are mainly found in blue-green algae and some species of bacteria.

Many years ago ideas were put forward to suggest that the prokaryotic cells were the first to emerge, but that nucleated cells were not just a simple mutation of the original cells. It was suggested that the eukaryotic cells were the product of a symbiotic union of several prokaryotic cells. At the time this was proposed the idea was largely dismissed by scientists. It is only in recent years that biologists like the American microbiologist Lynn Margulis have resuscitated the theory and added new theoretical weight to its premises.

It is of course impossible to prove a chain of events that occurred so long ago. All a historian or biologist can do is provide a theory that matches all the physical evidence and that follows a likely and logical path. Dr Margulis uses hereditary endosymbiosis (endo = within) to illustrate her theory. There is, for example, a plant called *Psychotria bacteriophila* which actually has a symbiotic bacterium in its seed. In this way the next generation receives all the usual hereditary material but also the symbiont from its parents; thus this is an instance of hereditary symbiosis.

Having established that an inherited symbiosis can exist, the propounders of the symbiotic theory of cell evolution turn to the eukaryotic cell itself. Within the eukaryotic cell there are tiny organelles as well as the nucleus. These cells also contain mitochondria in which the energy is produced for the benefit of the organism. Mitochondria have their own DNA, which is the genetic material of the cell, and this is separate from the nuclear DNA and unrelated to it. The mitochondrial DNA, like that of prokaryotes, is in simple strands, not in complex chromosomes as in the nuclei. These two parts of the cell, the mitochondria and the nucleus, therefore have the same basic genetic ability. It is abnormal in nature that two things should have the same function; this suggests that they evolved from two free-living organisms living in symbiosis. It was noted in the 1920s by J.E. Wallin how similarly the mitochondria resembles free-living bacteria that over a long period of time established an hereditary symbiosis with ancestral hosts ultimately evolving into animal cells, plant cells and cells that fit neither of these categories.

Dr Margulis and her colleagues also assert that other elements of the eukaryotic cell have arisen via symbiotic evolution. They name mitochondria, flagella and cilia and photosynthetic plastids as the symbiotic parts of the cell; they call this the serial endo-symbiosis theory. Only by experiment can proof be given to this theory. The symbiotic elements will have to be separated, cultured and reintroduced to their partners. At the moment nobody has been able even to grow any organelle from an eukaryotic cell. It does seem that the theory put forward answers some problems of cell evolution and fits with known data, and if it is indeed true evidence will eventually be found to back up the theory.

It may be seen from the above that symbiosis in its various forms is of vital importance to this whole planet. Some symbioses concern minor and very local associations whereas others produce an effect which concerns all plant and animal life in some way.

I do not claim that every existing symbiotic association is mentioned in this book; indeed it would be folly to do so. At this moment, new or previously undocumented species are being found, and this process of discovery is set to continue for a very long time. Even without as yet unknown examples of this kind of behaviour, however, several symbioses have been excluded on the grounds of repetition. As many species' associations with one another are almost identical, I have chosen to describe a typical relationship and any others which may differ to some important extent from the norm.

1 THE CORAL REEF

One of the most significant symbiotic associations is directly responsible for the formation of coral reefs. Partnerships between tiny unicellular algae and various marine animals together enable these coral formations to build. Coral is made up of the calcareous skeletons of millions upon millions of tiny marine organisms. They also deposit calcium during their lifetime. The lower cell layers of the organisms continuously lay down calcium so that the reef grows beneath this cell layer, building the coral continuously. The animal attaches itself to, or lives close by, the skeletons of its ancestors, and when it dies its calcium is added to the rest. These structures grow a little like the formation by slow dripping of stalagmites, gradually getting higher, longer and wider; but whereas stalagmites grow underground in still, undisturbed caves the erosive action of the sea causes pieces of coral to break off and thus slows down the growth rate.

There are many coral formations in our seas and they are all found in warm, clear, tropical, equatorial waters. They occur only within latitudes of 30° either side of the equator. Coral needs water that is shallow and free from pollution. Unless the tiny algae can receive sunlight through the waters above they are unable to photosynthesise. Any depths greater than 50 m would effectively filter away the necessary sunlight, and any pollution would do the same. Polluted particles reflect light in all directions and prevent it from going directly downwards to the algae.

Coral outcrops are categorised into three types: barrier reefs, atolls and fringing reefs. A reef is a long strip of coral usually running parallel to a coastline. Within its length islands and lagoons are formed; the most notable reef is the Great Barrier Reef off the east coast of Australia, which extends to some 2000 km. The number of individual skeletons that have gone to make this reef is so vast as to be unimaginable, as is the amount of calcium

A coral atoll at Bora
Bora Island, French
Polynesia, showing the
lagoon.

deposited by these animals during their lifetime. An atoll is usually
a horseshoe-shaped piece of coral. There are many examples of this
kind of formation, for example Bikini Atoll. A fringing reef is
smaller than the other type of reef and usually very close to a
shoreline, too small to have islands or lagoons.

There are many areas of coral which never attain sufficient
height to break the surface of the water. They lie beneath the
surface and provide shelter and food for many species of plant and
animal. Corals form an amazing variety of shapes and colours.
Some are solid-looking, rock-like formations; others are delicate,
branching, brightly-coloured structures. The animals and plants
that live in association with corals are also varied and numerous.
Several of those that form symbiotic associations with one another
will be discussed later in this chapter.

It could be said that any animal or plant that lives in close
proximity to corals lives in symbiosis with it. The coral enriches
the water with oxygen for these creatures, and they consume
plankton and debris to keep the waters clear enough for the coral
plants to survive. This type of association is probably crucial to
the survival of coral reefs and their attendant populations, but it
is rather unspecific. No one symbiotic act can be distinguished; it
is rather of general benefit to all that they should cohabit. The
specific relationship between the tiny plants and marine polyps is

15

the important association that enables coral structures to form, and that is the one on which we shall dwell.

The animal members of this partnership consist of several different species, but the one thing they have in common is the ability to deposit calcium. The plants are microscopic algae and are usually referred to as dinoflagellates. The coral-forming animals are filter feeders, sifting through the water and retaining animal plankton particles. They are strictly carnivorous in eating only animal matter. In the course of this filtering they take in the dinoflagellates along with their food. By some means, the tiny plants are separated out from all other food and pass into the body of the polyp. There they begin to photosynthesise. In some species symbionts are directly transmitted from generation to generation, which avoids the necessity for each generation to ingest the dinoflagellates.

As they photosynthesise the plants are able to provide their host with oxygen and carbohydrates surplus to their own needs. The dinoflagellates are also able to utilise their host's excretory products of carbon dioxide and nitrogen. They convert these waste products and make them usable again for the host. In this way the animals' waste is re-cycled. Some coral-forming animals have become so dependent on their tiny plant invaders that they stop feeding in the normal way and gain all their nutrients from the photosynthetic by-products obtained from the plants. These photosynthetic activities also increase the deposition of calcium in their host animals, which significantly increases reef formation.

The corals that are the strongest and most able to withstand wave erosion are the rock-like *Montastrea*, which are roundish and form the main part of the reef. The branching corals, that are by their nature more fragile, lie at greater depths, sheltered from the worst effects of the waves.

The ability of coral formations to grow in the face of constant battering from the sea has long fascinated man. T.F. Goreau, in his paper 'Problems of growth and calcium deposition in coral reefs', says that the wave action and other eroding forces mean that individual corals must grow at a faster rate than 40 cm per year, which is the maximum overall growth rate of the reef.

Coral reefs have dense populations of dinoflagellates, as for a coral outcrop to form many millions of animals must live within a confined area. In this situation the dinoflagellates help to oxygenate the water thus making the area more suitable for other forms of life. Most coral reefs are autotrophic, that is they produce more

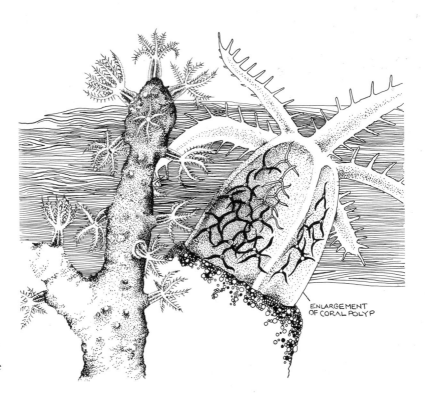

A typical coral polyp.
The enlarged section
on the right shows the
dinoflagellates.

ENLARGEMENT
OF CORAL POLYP

organic matter than they consume. The surplus organic matter is
able to support large colonies of other animals that live within or
close to the reef.

Clams and Algae

One of the most dramatic-looking inhabitants of the Indian and
Pacific ocean coral reefs is the Giant Clam (*Tridacna gigas*). This
enormous bivalve can grow to more than 1 m across and weights
of 200 kg are common. Popular legend has it that this creature
is aggressive and actively 'hunts' unwary skin divers. Many folk
stories tell of men having lost legs or being sucked under and
drowned by these clams. These tales originate from the animal's
self-protection mechanism of snapping shut its shell if attacked, or
suddenly shaded. The fact that the clam lives in shallow, clear,
tropical waters frequently infested with divers must add to the
weight of these stories. In truth, however, the clam is far from
aggressive; it is such an enormous, weighty creature that it can
hardly move at all and most certainly cannot 'hunt' anything.
The clam lives on the substratum of the ocean among corals, with

17

its shell open towards the surface of the water. The soft mantle spreads over the edge of the shell, becoming exposed to water and light. The mantle is brightly coloured and houses microscopic algae. The algae are unicellular and live underneath the ectoderm of the mantle. Here they are able to photosynthesise as they do in the coral animals. In the same way they give their hosts surplus carbohydrate and oxygen. This explains why the clam lives only in shallow, clear waters and usually among corals.

The algae are taken into the clams by the normal filter method. The animal filters the water, taking in all kinds of microscopic organisms. By an as yet unknown process, the clam is able to differentiate the algae from its normal food, and these selected organisms pass into the mantle cavity. If the algae population gets too large, some of them pass through into the clam's digestive system and are consumed, ensuring that a constant and reasonable population of algae is retained. This shows that the mechanisms that control the algae are very subtle and effective. They are sufficient to select the correct unicellular algae from the millions of organisms in its food, and also be able to maintain optimum population numbers within the mantle.

The benefits of this partnership to both species are considerable. The clam obtains oxygen and carbohydrates from the algae, and from the algae's point of view the clam is an ideal home, as the host

19

has no real enemies. Only man ventures to take clams, which are considered a delicacy; no other creature predates on it in significant numbers, as the rapid snapping together of the shell discourages attack.

Crabs and Sea Anemones

There is a crab called *Lybia tesselata* that lives within coral formations in the Indian Ocean near the Seychelles. It associates with small sea anemones and together they form a fairly close symbiotic relationship. The anemones are small, with tentacles around the 'mouth' end of their bodies. All along the tentacles are tiny stinging cells called nematocysts. These cells can stun or kill prey species which are then drawn into the anemone's body as food. *L. tesselata* uses these anemones as an additional weapon in its armoury. It already has the usual exoskeleton making it pretty well defended against predators, but feels the need for extra defences. The crab carries an anemone on each claw which it thrusts into the 'face' of any enemy that dares to attack. The stinging tentacles brush into the attacker and convince it that other prey would perhaps make an easier meal.

Lybia tesselata with anemones on each claw. The claws are no longer used for their normal function of feeding.

The anemone is held in place by small 'teeth' on the claw which grip it firmly. Some crabs keep anemones on their claws all the time whereas a few pick them up only when in danger from a predator. This relationship must be a very longstanding one in evolutionary terms, as the original function of the claw was to break up food and put it into the mouth. Such tasks are now impossible as the claw is taken up with the anemone. The function of the claw has passed onto the crab's two front legs which have become modified for this purpose, leaving the claws free to carry the anemone.

The two species both benefit from this relationship. The crab is better able to defend itself from predation and the anemone also gets protection from the crab, as few creatures attempt to eat the two species when they are working together. The anemone also gains from an extended habitat. Normally they are slow-moving creatures and consequently have a small range, but when living on a crab the anemone is able to move around a much wider habitat. This increases the feeding area and adds variety to the food sources available to the anemone.

Fish and Sea Anemones

There are several examples of symbiosis that involve different species of sea anemone, and probably the best known marine symbiosis falls into this category. It concerns anemones and damsel or clown fishes. These small, brightly coloured fish inhabit the warm, clear waters of tropical coral reefs. They have many different colours and patterns and associate with several species of anemone. The fish live among the stinging tentacles of the anemone and are completely unharmed by the poison. It was originally thought that the damsel had some natural chemical immunity to the toxin, but it is now believed that protection is gradually built up. When approaching the anemone for the first time the fish is very wary and only gently brushes into the tentacles. Slowly it becomes bolder, swimming through them until it can quite happily remain in the centre of the anemone in constant contact with the stinging cells. It is believed that the mucus layer surrounding the fish is modified by each minor sting until the protection is complete and the fish can touch the tentacles with impunity. This relationship built up by the fish with the anemone is highly individual, as the immunity is specific to one anemone only. If the fish moves onto another one it is just as vulnerable to the poison as any other fish.

Once the association has been built up, the fish lives in the

)verleaf: A damsel r clown fish living mong the tentacles of sea anemone.

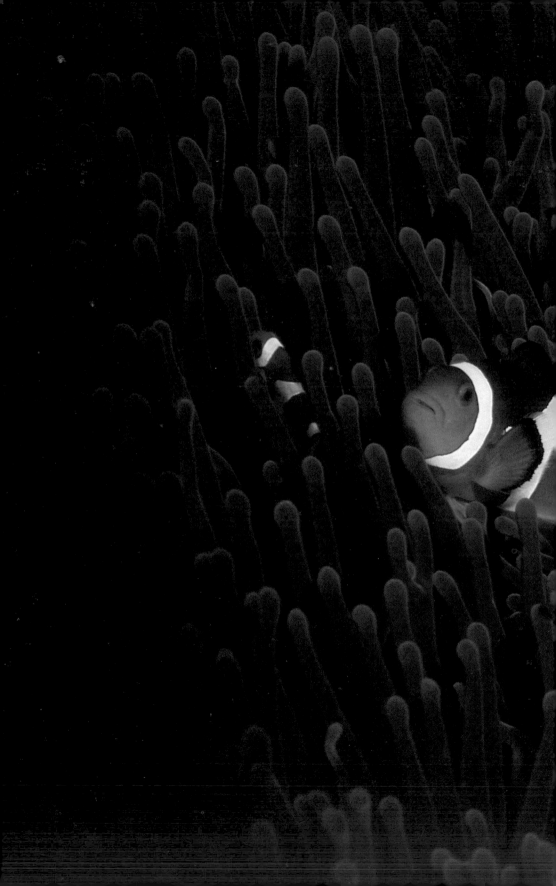

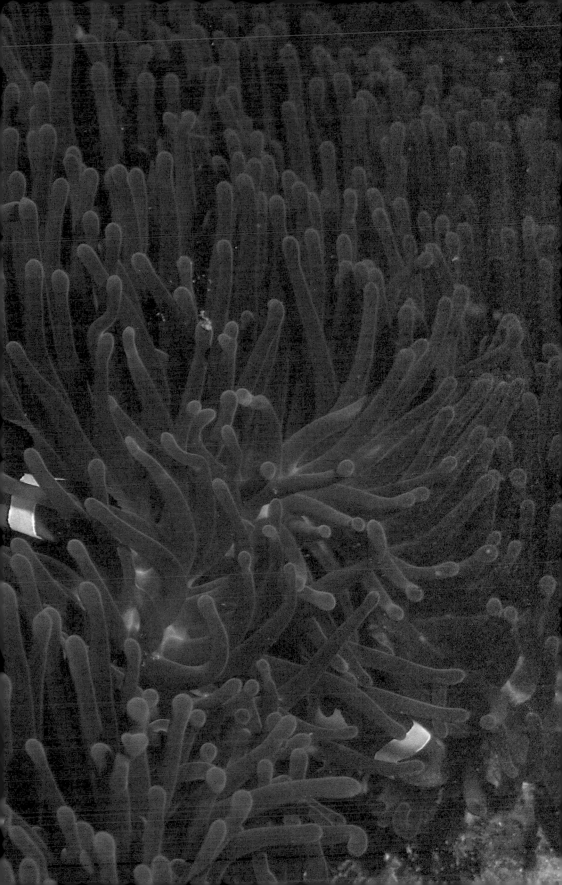

anemone only venturing out to catch its food. Quite often pairs of damsel fish live in one anemone, and occasionally large numbers of fish can be found in a single anemone. The anemone makes no attempt to eat its resident damsels but will happily consume any strange one that ventures too close. This would imply some discriminatory ability on the part of the anemone. However well protected against the toxin a fish may be it would seem reasonable that the anemone would try to eat it unless it were able to distinguish its own residents from strangers. Therefore, it is likely that there is some mechanism whereby the anemone and the fish are able to communicate, enabling the anemone to identify its own fish.

The relationship between the two species is very satisfactory for both parties. The fish is provided with a safe habitat, as anemones are a most unpalatable food source for other fish. While it remains within the tentacles it is most unlikely to be in danger from predators.

The anemone gains a partner that will remove debris and any infected tissue from its body, and a colourful lure to attract fishes to come close to its tentacles. The damsel is a highly favoured prey for many larger fish and attracts them to the anemone, which can then render them immobile with its sting and consume them, sharing any spilled food with the damsel.

Sea Urchins and Fish
One further example of a symbiotic association among coral reefs concerns the Hat Pin Sea Urchin (*Diadema* sp.) and two small fishes, the Shrimp Fish (*Aeoliscus strigatus*) and the Cling Fish (*Diademichthys lineatus*). This urchin is so named because of the resemblance of its long, thin, black, pointed spines to the pins used by Victorian ladies to prevent their hats slipping off.

The urchin is a slow-moving creature that lives among corals. It is inedible to other species, as the spines are sharp and inflexible, making them virtually inpenetrable to predators. This highly effective protection makes the urchin an ideal host for the two tiny fish that shelter there. The shrimp fish and cling fish live among the spines and clean up the urchin to keep it free from parasites and debris. This relationship must be of long evolutionary standing, as the fish have become physically adapted to this unusual habitat. They are very slim, elongated and swim vertically (noses down) so that they can move around among the closely grouped spines. The shrimp fish and cling fish are unrelated to one another and

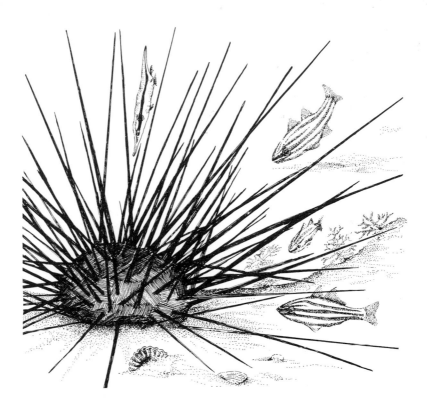

A hat pin sea urchin with a shrimp fish (left) and cling fish which have adapted to living within the spines.

must have evolved separately, but because they choose the same environment they have evolved to be physically and behaviourally very similar.

This relationship is of great mutual benefit as the fish have a near perfect, safe home and the urchin has a permanent cleaning service.

Shrimps

Other coral species act as 'cleaners', and the shrimp *Periclimenes petersoni* is one of these. Although Chapter 3 deals specifically with the 'cleaners', this shrimp will be discussed here, as it is an important part of the life of the coral reef. It lives among the corals around the Bahamas in association with an anemone called *Fartholomea annulata*. It is not fully understood what the reason for this association is, but the shrimp is almost always found clinging to this particular anemone.

P. petersoni has a transparent body with white stripes and violet spots, and it has very long antennae, longer than the rest of its body. While waiting for customers the shrimp waves its antennae in order to attract a client. This shrimp seems to be particularly trusted as many fish will allow it into their gill cavities and their

A cleaner shrimp cleaning the gills of a copper-banded butterfly fish.

mouths. It crawls all over the client cleaning off necrotic matter and parasites. The client waits patiently while being groomed, and even allows the shrimp to bite in order to remove parasites from under the 'skin'. Other species also act as cleaners.

The coral reef is a very important marine phenomenon, as it produces more nutrients than it requires for its own survival. This helps to nourish the waters around the reef and so maintain the life of adjacent waters. Corals can produce massive structures the size of the Great Barrier Reef and have made a considerable contribution to the face of the world. The fact that a coral formation can grow in the face of natural sea erosion is truly amazing, as is the beauty and variety of the corals themselves. The coral reef is also of considerable importance as a habitat for countless species of plant, fish and animal, some of which have been mentioned in the context of symbiotic associations. It is fascinating to think that without the photosynthesis of dinoflagellates within the coral polyps these massive and beautiful reefs could not have been formed, and the habitat for all the coral-dwelling creatures would not exist.

2 MILKERS

This chapter is concerned with several species that effectively 'milk' other species for a foodstuff that they desire. The process is generally the same for each animal; the milker strokes the other creature, thus stimulating it to ooze a substance usually called honeydew. This substance is neither honey nor dew, but it is a sweet solution and its name is probably derived from this.

Honeydew is thought to be the 'manna from heaven' referred to in the Bible. Mealy bugs and aphids exude this sticky substance over the leaves of the bushes on which they live, and then the fierce heat of the sun dehydrates it, and a white crystalline formation is left. This glistens in the sun like frost on leaves and is often collected by native peoples as food. The collector can obtain a kilogram of honeydew a day and sell it as a delicacy in the local markets.

Three milking relationships are described in this chapter, which are all quite different from one another and indicate the range of these associations. There are some other relationships which do not vary much from the ones detailed.

Aphids and Ants
The green, white and black flies that plague gardeners are types of aphid. These creatures have developed mouthparts that enable them to pierce the outer cells of plant stems so that they may tap the sugary juices of the plant. This is why gardeners and horticulturists dislike aphids, as they consume the plant's nutrients and consequently hinder its growth and development. They can also transmit disease. Aphids suck the highly concentrated sugar solution from the plant and extract the tiny amount of useful nutrients that are contained in this solution. Because such a small amount of the liquid is useful the aphid has to consume large quantities of the plant juices and has therefore to get rid of considerable amounts of

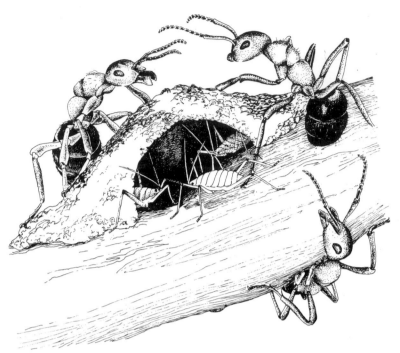

Ants construct shelters from mud for the protection of their domesticated aphid populations.

unwanted sugar and water. This is the sticky waste substance called honeydew, which is exuded from the aphid's anus.

Wherever colonies of aphids are found ants of various species often abound as well. This is no coincidence, as the ants are interested in the aphids' honeydew. They find this substance almost irresistible, although it is not a very useful food. The ants approach the aphid from the rear and stroke its hind end; this stimulates the aphid to release honeydew, which the ants lap up.

Various species of ant have extended this basic relationship and have developed different methods of colonising or domesticating the aphids. This adds the element of symbiosis to an association that would otherwise be only one-sided. In one species, the ants take fine earth up to the leaves and stems of plants and, using their own saliva, cement together tiny shelters, shaped like mud huts, for their aphid partners. These shelters help to protect the aphids from severe weather and to some extent from predators.

Other ant species will go out searching for aphid colonies and gently bring them back to a convenient plant, close to the ant nest. This makes the aphids constantly 'on tap' for milking by the ants. Some ants are left on duty to protect the aphids from predation.

As the ants have a vicious bite, their presence keeps away many species of aphid predator.

Some ants will round up local populations of aphids at the end of the day, in much the same way that a sheepdog herds sheep. The ants then take their aphids down into the nest for protection from predators. In the morning the aphids are escorted to the required plant for another day's feeding and milking.

Ants have been observed collecting tiny aphid eggs and gently taking them into storage chambers in the nest to that they may overwinter with some protection. Normally these vulnerable eggs have to survive the rigours of winter, stuck onto the leaves and stems of plants. Many are lost through being eaten, and from simply being blown off. They are much more likely to hatch if protected from these dangers inside the nest. In spring the ants return the eggs onto suitable plants so that they can hatch in the normal way. Once the aphids have hatched they are protected and milked as usual.

It is interesting to wonder how this complex series of actions came about, and how the first ants came upon the obvious advantages of this association. Most actions are a result of incentive and as ants seem inordinately fond of honeydew this is a powerful driving force which makes them search around for ways of keeping a constant and large population of aphids available at all times. Honeydew is not an essential food for ants and the reason why ants and other species find it so attractive is probably analogous to the 'sticky bun' craving of many human beings. It is not only mealy bugs and aphids that produce honeydew; some larval caterpillars produce a very similar substance, and the ubiquitous ants are always around to take advantage of the situation.

Butterflies and Ants

The common Imperial Blue Butterfly (*Jalmenus evagoras*) lives in eastern Australia, in tropical and sub-tropical regions. It feeds on a shrub-like species of acacia called the Black Wattle. The female butterfly lays her eggs on the branches of the wattle; the eggs, which are laid in clumps, are white and covered in urchin-like spines. Large colonies of Meat Ants (*Irido myrmex*) also live near the wattle bushes and they surround the eggs waiting for them to hatch. When the larvae emerge they are followed around by the ants, not being molested in any way; the ants are there to guard and protect the caterpillars from predators. These ants have a vicious bite which they are not afraid to use on any insect trying to

Above: A red ant milking honeydew from an aphid.

Left: An Imperial blue butterfly larva being milked by a meat ant.

Right: Red ants farming aphids.

feed on the larvae. This bite is sufficient to cause pain to humans and other large animals and must be a considerable disincentive to smaller creatures.

At this time the association is totally one-sided; the larvae are protected but the ants receive nothing in return. It is not until the caterpillars have grown considerably larger that they are able to exude the honeydew that the ants desire. When they have reached this stage the ants milk them by stroking with their antennae the rear part of the caterpillar.

Eventually the caterpillar pupates, forming a chrysalis. Once again the ants guard their charges fiercely for no gain, as the chrysalis does not feed or exude honeydew. When the butterfly emerges, there is an ironic twist to the story. The creature that has been so well protected all its life by the ants is suddenly turned on by its protectors. The butterfly has consequently developed the ability to fly immediately on emerging from the chrysalis, so as to enable it to escape the attacking ants. Normally a newly emerged butterfly remains still for a few minutes to allow time for its wings to fill with blood ready for flying. If for any reason the butterfly is unable to fly away the ants attack and kill it.

This behaviour does not imply any cogent thought processes on the part of the ants, rather that they act instinctively; but why have the ants not learned that their honeydew supply ends with the pupation of the caterpillar? Other ants have been able to work out much more complex cycles via trial and error (see the previous example of ants collecting and storing aphid eggs). It seems strange that over all the years that this process has been performed none of the ants have learned that they are wasting their time and efforts guarding the chrysalis. Perhaps it is the simple greed of the ants for honeydew that drives them to wait in vain for the larvae to re-emerge. Although the ants do not seem to be able to work out that the caterpillar has changed into a butterfly, the Imperial Blue on the other hand has become able to fly immediately after emerging. This ability is presumably a result of evolutionary pressure over the years that this association has been going on.

There is another ant/butterfly relationship that displays one of the more extreme symbioses. It concerns the Large Blue Butterfly (*Maculinea arion*) and the common Red Ants (*Myrmica scabrinoides* and *M. laevonoides*). The large blue was native to Great Britain, but has now been declared extinct there, although it is still found in many parts of the European and Asian continents. Until 1915 the life history of the large blue was unknown. The mystery about the butterfly's life cycle surrounded the seeming disappearance of the caterpillar from the third moult until the adult butterfly appeared. No chrysalids or caterpillars after this moult were ever seen, which left a considerable section of the life cycle a complete mystery. Several experiments were conducted on the caterpillars in an attempt to rear butterflies in captivity and explain the missing stages in their life cycle, but at the same stage (the third moult) the caterpillars became listless, refused their food and eventually died.

In 1915 two experts (Frowhawk and Chapman) saw some wild thyme, on which the large blue lived, growing on an anthill in Cornwall. They pulled up the plant to look inside the ant nest and among the ants they found a chrysalis that was unfamilar to them. As they were able to identify the chrysalises of all the endemic species of butterfly, the two men assumed that this could be the hitherto unseen large blue chrysalis. They then began to conduct experiments with the caterpillar and ants and were eventually able to piece together the missing elements in the life story.

They discovered that after the third moult the caterpillar descends from the thyme plant and wanders around in an apparently pointless manner until, by chance, it comes upon a red ant. The ant moves about agitatedly and in this highly excited state moves to a gland on the caterpillar's seventh abdominal segment which it then strokes with its antennae. This gland exudes a sweet sticky substance very similar to the honeydew produced by the aphids, caterpillars and mealy bugs already mentioned. The ant laps up the honeydew with great enthusiasm, and after drinking the substance it picks up the caterpillar and takes it down into the nest.

Inside the nest the caterpillar is treated as if it were a royal visitor. In exchange for a constant supply of food it allows the ants to 'milk' it of the sugar solution. The ants are so concerned for the well-being of the caterpillar that they even feed it on their own young grubs. In this way the visitor is cared for until it hibernates, pupates and forms a chrysalis. During pupation it is protected from the dangers of the outside world by the ant nest and remains unmolested by the ants themselves.

When the chrysalis hatches the butterfly's wings remain folded until it has had time to walk along the narrow passageways of the ant nest to the open air, before its wings fill with blood and straighten out. In this way the cycle of life begins again; the butterfly lives its life, mates, lays eggs, which hatch, and the caterpillars proceed in the same way with the ants' co-operation until the next generation of butterflies is produced.

All attempts to rear the large blue caterpillar artificially without ant grubs, during the vital stages of its development, have failed to produce adult butterflies (because the caterpillar will not pupate). It is therefore assumed that some nutrient within the grubs is essential for the caterpillar to develop and pupate.

The relationship between the ants and caterpillars is one of the closer examples of symbiosis, as one species is totally dependent on the other for its survival. The association is somewhat one-sided, as the honeydew is not essential to the ants but they go to considerable trouble to get and keep the caterpillars.

The relationships between 'milkers' and honeydew producers is so extraordinary and highly developed that they must be of considerable evolutionary standing. It is unusual for such close relationships to develop when only one element in the partnership actually benefits in any real and essential way.

A large blue butterfly larva about to be carried into the red ant nest.

3 CLEANERS

Cleaning symbioses are found in the sea, in freshwater, on land and in the air, but the greatest number of examples concern marine species. Some creatures are physiologically unable to keep themselves clean and free from parasites, as they are shaped in such a way that it is impossible for them to remove any foreign matter from their bodies. Fish are obvious examples of this. They are not able to bend in two, and although some can reach their tails with their mouths it is not possible for them to reach any other part of their bodies. There are other animals that need assistance with cleaning, such as hippopotami. Their mouths are so huge and ungainly that even if they could reach the parasites they would be totally incapable of removing them; the same applies to crocodiles, alligators, some iguanas, elephants, rhinoceroses and turtles.

Other animals are able to groom most of their bodies but find small areas difficult. These creatures often engage in mutual grooming with their own kind; apes and monkeys are good examples.

It is essential for all creatures to have some method of keeping themselves clean and free from parasites. If they do not, they will probably fall ill from infected wounds or the effects of disease and blood loss from parasites. For those species that are unable to clean themselves it is obviously vital to find some other animal to perform this cleaning function. This chapter is concerned with describing some typical examples of cleaning symbioses as well as the more extraordinary ones.

The advent of sophisticated diving equipment, which allows man to remain underwater for considerable lengths of time, has enabled us to study marine cleaning symbioses in considerable detail. Cleaners are found in all the oceans, and sometimes the associations are between closely related species but they are often between distant ones. Cleaner fishes are small, agile and brightly

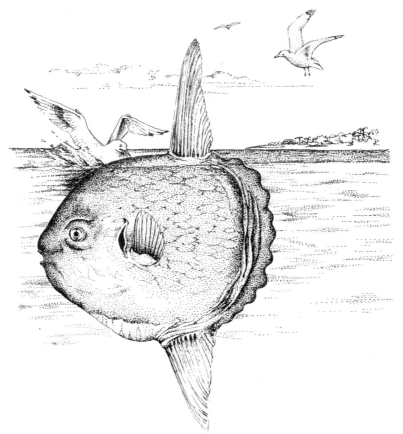

Seagulls clean the parts of the ocean sunfish which protrude above the water. This fish also floats on its side for periods which enables the birds to reach more of its surface.

coloured, thus they are easily recognised by the 'client' species and nimble enough to perform their tasks. Several cleaners advertise their services by taking up strange postures, waving antennae and 'dancing'. Cleaners are usually trusted implicitly by their clients and are allowed into the mouths and even gill cavities. The parasites that they remove provide them with a large part of their diet, but no species has been shown to be able to live totally from parasites removed from other animals. The relationship between cleaner and host varies considerably from casual and accidental to an association quite deliberately brought about by either party and repeated over and over again.

It is not only fish that act as cleaners; shrimps of various species also perform this service (*see* Chapter 1). At least two species of crab have been seen to remove parasites from land mammals and marine iguanas. Various species of seagull are known to pick parasites from the Ocean Sunfish (*Mola mola*).

The Nassau grouper
and its goby cleaners.
The host is particularly
protective towards the
goby.

Fish

The vast majority of cleaners are fish; at least 45 species are known cleaners and there may well be many more. Fish that are habitually cleaned often have to modify their usual behaviour to allow the cleaners to do their work. It is not normal for aggressive species like shark, barracuda and moray eels to allow small fish to swim safely near them. With known cleaner species, however, these and other fish change their attitude completely and allow the cleaners all over their bodies without displaying any ferocity towards them. The clients will slow down or stop completely (unusual behaviour for most fish, as they usually move all the time), open and close their mouths and gill covers and assume awkward-looking postures to help the cleaners. It is quite possible that some species have become extinct because of an inability to establish a cleaning symbiosis. So many individuals could have fallen foul of ectoparasites, fungi and bacteria that the population was made inviable.

Some fishes change colour while being cleaned. Black Surgeon Fish (*Acanthurus achilles*) go from black to blue when they are being cleaned by *Labroides dimidiatus*. The Goatfish (*Parupeneus trifasciatus*) changes from pale brown to pink while the same

cleaner picks it over for parasites. The reason for this behaviour is not understood at the moment, although some work is being done on the subject.

Fishes being groomed guard their cleaners against danger by warning them of the approach of predators. The Nassau Grouper (*Epinephelus striatus*) when cleaned by gobies warns its cleaner by suddenly closing its mouth, leaving only a small gap to allow the goby to escape. Even if the grouper is in imminent danger itself it takes time to warn the goby. This shows the regard that the client feels for its cleaner and the service that it performs.

A cherry barb showing the white fungal infection which has occurred because the fish cannot clean itself and has not received the services of a cleaning species.

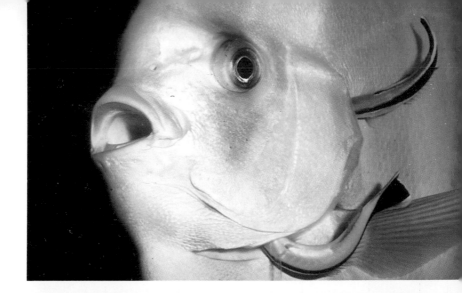

Cleaner wrasse cleaning the gill cavities of a batfish.

The client fish (here a coral trout) will even allow the cleaner inside its mouth, which calls for a great degree of trust on the part of both fish.

Several species of cleaner set up cleaning stations in one particular place. The local fish soon realise where the cleaner is located and will visit it whenever they require cleaning. Quite astonishing numbers of fish are cleaned in this way: not only territorial species that would normally be found in the area but also migratory ones which have gone out of their way to visit these stations. Client fish will patiently wait their turn to be cleaned, and even form orderly queues.

Quite a considerable amount of observational and experimental work has been done on these cleaning stations. Limbaugh, for example, discovered that over three hundred fish can be cleaned by a single Senorita Fish (*Oxyjulis californica*) in a six-hour period. These fish go back to the same cleaner every few days for another session and this enables them to remain in peak condition.

Limbaugh also did some experiments in waters off the Bahamas. He removed all the cleaner fish from one locality and observed the effects on the species normally found there. Within two days the numbers of fish were severely reduced and within two weeks almost all the territorial fish had disappeared. Those that remained had developed the fuzzy marks that are an indication of fungal infection. It had been shown in previous experiments that the introduction of cleaners into an aquarium infected by fungi can restore its inhabitants to health.

From the above the value of cleaning symbioses in the marine habitat can easily been seen. Without the work of all the cleaners of the oceans, the effects of parasites, fungi and injury would kill many more species than they do already. The Senorita Fish (*Oxyjulis californica*) is an example of a typical cleaner. It is of the wrasse family and lives off the coast of California. It is an active,

The cleaner wrasse attending to a surfperch. The client fish queue up in an orderly fashion for the cleaning service.

small, cigar-shaped fish that the local people call the senorita because of its cleaning habits. Its client fishes include the Topsmelt (*Atherinops affinis*), Black Sea Bass (*Stereolepis gigas*), Opaleye (*Girella nigricans*), Blacksmith Fish (*Chromis punctipinnis*) and many more. These fish are almost all much larger than the cleaner and would normally prey on wrasses of the senorita's size. They do not attack the senorita, however, but wait patiently until it is their turn to be cleaned, hold themselves still and often in the most peculiar postures while being attended to. The fish in the area of coast that the senorita fish inhabits are especially troubled by fungal infections, and removal of the white growths caused by the fungi is the cleaner's most important function. The cleaning phenomenon has been observed for many years to the extent that one species is popularly called the Cleaner Fish or Wrasse (*Labroides dimidiatus*). It is a small, slim fish with a cyan-coloured body, striped with darker blue or black. The cleaner fish goes one stage further than the senorita in that it actively attracts clients by 'dancing'. It swims in a vertical position, head downwards, and undulates its body from side to side. This is a most unusual posture

The blenny or false cleaner mimicks the vertical dancing behaviour which the cleaner wrasse uses to attract its clients.

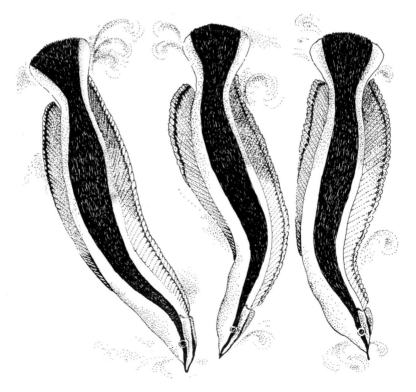

for fish, as they usually swim horizontally to the sea bed. This 'dancing' makes the cleaner noticeable to even the most myopic fish, and it has become the cleaner's trade mark.

Clients line up, as for the senorita, until it is their turn to be cleaned, and also allow the little fish to enter their mouths and gill cavities unharmed. The contents of various species' stomachs have been examined to assess the quantity of cleaner fish that are eaten, both by fish that are known clients and others. It has been found that very few cleaners are consumed by any species, although fish of similar size make up the bulk of the diet. So few cleaner fish are eaten that it seems probable that the small number that are are taken accidently by absent-minded clients rather than actively predated upon.

This cleaning behaviour must have been going on for some considerable time in evolutionary terms as the cleaner fish has acquired a copy-cat species. The Blenny (*Aspidontus taeniatus*) has learned to take advantage of the trust given to the true cleaner. It mimics *L. dimidiatus'* behaviour by copying the 'dancing' technique, and it also resembles the cleaner in shape and colour. Because of its behaviour and appearance, the client fish are fooled into thinking that the blenny is a cleaner. They allow it to approach in the normal way, but instead of removing parasites from the client the blenny bites chunks out of the host's fins. This activity has given rise to the blenny's popular name of the 'false cleaner'. This subterfuge usually works the first time the client meets a blenny, but it soon learns to distinguish the two species. The clients can identify the true cleaner because it always approaches from the front whereas the blenny approaches from the rear, desiring a piece of tail fin.

Another method of cleaning is practised by the family of fishes called the Sucker Fish (Remora). These fish are equipped with large oval suckers on the top of their heads with which to attach themselves to their clients. The strength of this suction is so great that fishermen have attached live sucker fish to their lines, and when the sucker sticks itself to another fish the man is able to pull them both onto his boat without the suction breaking. This attachment is so powerful that huge fish can be landed by this method.

In their natural situation sucker fish fix onto the under surface of sharks, rays or turtles and remain there until the host stops to feed. Most of these host species are messy eaters, biting lumps off their prey and spilling a great deal of food in the process. When it disconnects itself from its host the remora shares in the feast and

A sucker fish attached to the carapace of a turtle. The sucker is located on the back of the head.

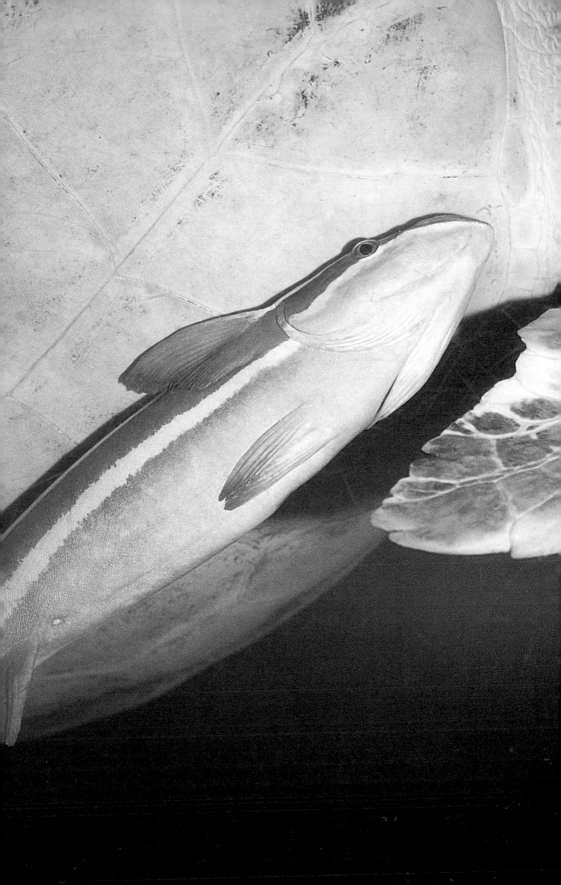

only attaches itself again when the host is finished and ready to move off. While the host is on the move the sucker fish travels all over its body removing parasites and necrotic matter. In this way the remora have a comfortable life, as they are most unlikely to be attacked while attached to the bodies of large, predatory species. They also have a constant food supply in the form of parasites and the scraps from their hosts. The host fishes benefit from the usual cleaning service, but instead of returning to a cleaning station they carry their cleaner around with them all the time. There are some ten different species of sucker fish varying in size from 4 to nearly 7 cm. Each variety seems to prefer a different host species.

The behaviour of these remoras has been known and recorded for many years. The ancient Greeks believed that the fish would cling to the bottom of boats, and that if enough of them did so they would slow down or stop altogether the boat's passage through the water.

Crabs and Iguanas
The Red Rock Crab (*Grapsus grapsus*) cleans other species, particularly the Marine Iguana (*Amblyrhynchus cristatus*) of the Galapagos archipelago. This iguana is found on several of the Galapagos islands in the Pacific Ocean, some 950 km (600 miles) from the South American mainland. These islands are close to the equator, volcanic and very barren with little plant life. The marine iguana lives on the shoreline grazing on seaweed at the water's edge, and basking in the sun on the warm rocks. It is a reptile and has to conserve body heat by soaking up the sun and remaining still for long periods. This sedentary life allows the red rock crab to scuttle all over the iguana's body while it rests. The crab removes any necrotic matter and parasites. Iguanas are particularly bothered by ticks, and the crab takes hold of these in its claws and tugs away until they come off. The iguana tolerates this painful process and even turns over to allow the crab to 'do' its underside. The crabs seem to be fond of eating these ticks and often spend long periods looking for what seems to be only a small quantity of food.

The iguana obviously benefits from this cleaning service as ticks can seriously debilitate an animal, and they are ill equipped for cleaning themselves. The crab gains a certain amount of much favoured food for its trouble. The crab is not in any real danger from this particular iguana, as it eats only seaweed and not animal foods. This relationship is fairly casual, but there are benefits for both parties.

Whales and Birds

One of the more extraordinary relationships between marine species is the association of the Sperm Whale (*Physeter catodon*) and a bird called the Grey Phalarope (*Phalaropus fulicarius*). The sperm whale is the largest of the toothed whales and is easily recognised by its enormous squarish head which takes up roughly one-third of the whale's body length. It feeds largely on the giant squid and is often found to have huge scars which it receives when trying to capture this prey. This animal, like other species of whale, is unable to clean its own body but is often troubled by lice-like parasitic crustaceans. These tiny creatures live in the crevices of the whale's body, because if they were to cling to the smoother surfaces of the whale they would be swept off by the pressure of water that flows past as it swims. By remaining in these crevices the crustaceans are able to avoid the main stream of water and remain fixed.

As the whale is a mammal it has to come to the surface to breathe. This breathing process takes several seconds as the whale first expels the used air, and then breathes in through its blowhole. In this brief time the phalarope lands on the whale's back and pecks off crustacean parasites and any infected wound tissue. The bird can only clean those parts of the whale's body that are close to or above the surface of the water, but this constitutes quite a considerable amount of the mammal's surface (*see* p. 50).

This phenomenon seems at first to be almost unbelievable; how does a bird first find the whale (which might be hundreds of miles from land) and secondly how does it time the cleaning process so accurately as to be able to land, clean and fly away before the whale submerges again? No conclusive work has been done on this relationship, but as this particular whale travels around in schools of up to a hundred individuals the phalarope could be fooled into thinking that they are a type of mobile beach. The total area of so many whale backs moving together could well give that impression. At any one time several whales must be on the surface to breathe and this could establish sufficient continuity to make the bird believe that it is following a solid structure. It is also probable that these birds follow the schools of whales as they are assured of a rich and plentiful feeding ground if they do.

The phalarope lives on an exclusively crustacean diet and usually feeds off the beaches of Africa and South America as well as skimming crustaceans from the surface of the sea.

This then is another example of a loosely symbiotic relationship. Both parties benefit from the association, but neither are in any

Overleaf: A red rock crab among a colony of marine iguanas. The crab provides a cleaning service for the iguana.

45

way dependent on the other. The whale receives a partial cleaning service and the bird an abundant, if unusual, food source.

Lizards and Birds

The Tuatara Lizard (*Sphenodon punctatus*) also has a loosely symbiotic relationship with two birds, the Diving Petrel (*Pelecanoides* sp.) and the Sooty Shearwater (*Puffinus griseus*). The tuatara is often referred to as a living fossil as it is the only remaining example of the reptilian order Rhynchocephalia; the rest of this group of animals have been extinct for about a hundred million years. This species is an example of an animal that existed in the mesozoic period, and it has lived almost without change for two hundred million years. The tuatara has seemingly been passed over by evolution, probably because of its isolation on one or two islands off New Zealand. This remoteness of habitat has prevented it coming into competition with any predator, and consequently it has not needed the kinds of modification to body or behaviour that more widely distributed species have.

The tuatara lizard shares the burrow inhabited by the sooty shearwater.

48

The common wall
lizard removes from a
gull's nest the parasites
which would otherwise
trouble the chick.

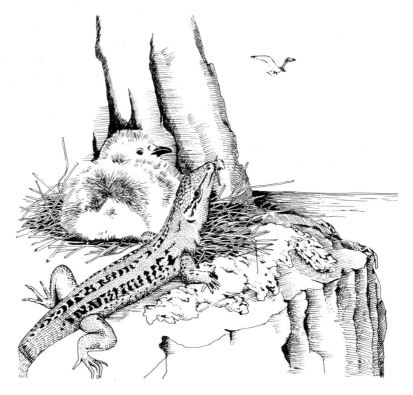

This lizard has a crest of elongated movable plates along its neck and back which has given rise to its popular name, the tuatara, which is the Maori word for 'spine bearer'. It probably does not reproduce until it is about twenty years old and lays its eggs in a burrow. The eggs remain in the nest for up to sixteen months which is a very long time for a reptile or, for that matter, any bird. It is thought that the lizard's life span might be as much as three hundred years. The tuatara can excavate burrows from the sandy soil on which it lives, but seems to prefer to live in ones already made by either a diving petrel or a sooty shearwater. These birds associate in colonies and their burrows are quite close together, which gives the lizard a measure of safety, as the birds have a noisy early warning system when danger threatens. From the birds' point of view, it is possible that the lizard may also protect eggs and nestlings from nest thieves in addition to its proven activities of cleaning the burrows of parasitic insects.

Another lizard known to associate with birds is the Common Wall Lizard (*Lacerta muralis*). Large colonies of several species of gull nest in the islands of the eastern Mediterranean, usually in crevices along the rocky shorelines. They nest during the hot, dry

49

summers and do not need to provide their young with any real weatherproofing; a few twigs or some soft nest material is put into a crevice or hollow formed naturally by the rocks.

The weather in this part of the world during the gulls' nesting time is hot, and consequently the nests are infested with a great number and variety of parasitic insects, which trouble both the parents and young. The gulls are unable to rid their nests of these parasites, and instead rely upon the wall lizard to clean up for them. The wall lizard is so named because of its ability to climb the sheer faces of walls or rocks. This agility and surefootedness allows it to reach even the most out-of-the-way nest. The lizards are able to catch and eat all the parasites and are often found to turn over twigs and other nest materials to look for insects.

The gulls seem to be aware of the benefits of this cleaning service, as they leave the tiny lizard unmolested. Reptiles of their size would normally form part of the gulls' diet, and so the birds must be able to differentiate the wall lizard from other species. The relationship formed by the gulls and lizards is again not essential; nor is the bond particularly strong. It is, however, of considerable benefit to both species to associate in the way that they do.

Above: Grey phalarop cleaning the moving backs of sperm whale far out to sea.

Right: Red-billed oxpeckers pecking fo parasites and cleaning infected tissue on the back of a buffalo in Kenya.

Birds and Herbivorous Mammals

The relationship between the Cattle Egret (*Ardeola ibis*) and various breeds of gazelle, zebra, antelope, giraffe and elephant is one of the symbiotic associations known to most people. The small white bird is seen sitting on the backs of these hosts in numerous wildlife films from Africa. The cattle egret's way of life must be highly successful as it is one of the few species that has actually extended its territory in recent years. In the late nineteenth century the birds were found to have colonised the northern coasts of South America, presumably by flying there across the Atlantic from their original home in Africa. Numbers in America increased rapidly and they now inhabit large areas of North and South America, and are still extending their range. Some cattle egrets have recently been found in Australia, and they may well eventually colonise that continent too. While spreading their range the egrets have taken with them their symbiotic habits, even though the host species have varied according to the local available herbivores.

The cattle egret is a small species of heron and has a similar lifestyle to the oxpecker and cow bird which will be discussed later. The egret, however, does not spend as much time as these other two species actually on the body of its host. The egret is normally to be found hopping around the feet of any type of herbivore. Its diet consists mainly of grasshoppers, which are superbly camouflaged and difficult to catch. The egret waits until its prey is disturbed by the movement of the host and catches the grasshoppers while they are in the air and easier to see. The egret does, however, spend some of its time on the body of its host, and while it is there performs the usual parasite-removing service as well as cleaning up any wounds the animal might have. The egret also provides its host with a sensitive early warning system. When predators approach the little bird hops up and down on the host's back, calling and flapping its wings. If the host is too slow in responding to the danger, the egret hops onto its head and drums away with its beak on the host's skull. This usually gets through to even the slowest animal, which then takes avoiding action. This relationship is fairly casual in that one bird does not usually stay with one host, but as both reap benefit the association is successful.

A relationship between a bird and the same types of herbivore, which differs slightly from that of the egret, is that between the Yellow- and Red-billed Oxpeckers (*Buphagus africanus* and *B. erythrorhynchus*) and the grazing animals of Africa. These two species of oxpecker inhabit different areas of Africa; between them

Cattle egrets hop round the feet of zebra and other African grazing animals.

their ranges cover the whole continent with the exception of the Sahara desert. They spend almost all of their lives clinging onto their hosts, usually in the woodpecker position—vertical with their heads raised. The only time they leave the host is during nesting or when startled. They use the host as a complete 'home'. They sunbathe, court, eat, sleep and mate while on the host and even remove hairs from mane and tail as nest material. The oxpeckers feed only while on the host, making no attempt to catch insects when on the ground. The diet includes flies, parasites of all kinds (especially ticks), and blood and tissue from the host's wounds. The oxpeckers' diet is highly specialised as they require the blood from engorged ticks rather than the insect itself.

The oxpecker has evolved physically to allow it to cling onto an ever moving host. Its claws are very sharp and curved to aid gripping, and its tail is long and stiff, which helps with the bird's balance. Its beak is unusually flat as well, which assists the bird in gripping onto the slippery bodies of ticks so that they can be pulled off the host.

Oxpeckers act in the same way as cattle egrets to provide an alarm system. They too have been observed to drum on the host's

53

head if it is slow in moving away from danger. The oxpecker is more dependent than the egret on its host, as it has to have the blood of engorged insects for its diet whereas the egret can catch any kind of insect. In areas where wild herbivores have diminished in number the oxpecker has readily moved onto domestic cattle to replace the normal hosts. Native farmers actively encourage the oxpecker as they are aware of the benefit accruing to their stock from the work of these birds.

This relationship is therefore a more extreme version of the egret/herbivore one. In this case the bird is much more dependent

Normally small numbers of oxpeckers attend the host, but occasionally the proportions become ridiculous!

54

on its host for whom it provides an improved cleaning service, spending almost its whole life at work.

Birds and Crocodiles

Another example of a cleaner bird is the Egyptian Plover (*Pluvianus aegyptius*) which grooms the Nile Crocodile (*Crocodilus niloticus*). This is one of the largest species of crocodile, living across Africa from Senegal to Natal in or by rivers and lakes. The waters in which the crocodile lives are warm, muddy and infested with parasitic leeches and other pests. The crocodile cannot avoid hosting these leeches as it spends a great deal of time in the water. Crocodiles are totally incapable of self-grooming, as their mouths are huge and full of teeth for catching prey rather than removing tiny parasites. They lie on the river banks for long periods with their mouths gaping open to encourage evaporation, and consequently the cooling of their bodies. This has allowed them to develop a symbiotic association with the plover. The leeches have difficulty obtaining a hold on the extremely thick, tough skin on the crocodile's body, so they tend to congregate in and around the mouth where the skin is softer and thinner. As the crocodile basks in the sunshine the plover is able to hop in and out of its open mouth to remove the leeches, any food debris stuck between the teeth, and scar tissue.

It is difficult to establish whether the crocodile is aware of the service performed by the plover, but it makes no attempt to kill or even scare the bird away. Crocodiles do eat birds of the plover's size and it would seem reasonable to assume that they would eat the plover if they were not happy with the arrangement. Crocodiles usually bask in this way during the hottest part of the day and after feeding; both these facts may considerably increase the life expectancy of the plover as the crocodile may be just too hot and full to be bothered by the antics of the bird.

The crocodile benefits in the usual way as host to a cleaner and also has this alert little bird as a sentinel. The plover gains a reliable food source and the 'protection' of one of the most aggressive and feared species of the whole continent.

Fish and Hippopotami

A similar arrangement is the association between the Hippopotamus (*Hippopotamus amphibius*) and a carp-like fish (*Labeo velifer*). The hippopotamus is a large, bulky animal which weighs about 1500 kg. It spends its life in similar rivers to the crocodile in central

Africa. It is well adapted to living in water as its eyes, nostrils and ears are all in a straight line on the top of its head, allowing the hippopotamus to submerge all but the very top of its head to escape the scorching sun, and yet keep all three sensory organs functioning. The hippopotamus lives in schools of anything from ten to a hundred animals in close proximity to one another. They excrete only in the water, and the action of all those massive bodies churns up the mud and excreta. This rich, muddy water provides an ideal habitat for many species of parasite and a small carp-like fish. The Africans call this little fish 'cattle egret of the water' as it performs a similar function to the egret (*see* p. 58).

The fish swims all over the hippopotamus' body removing parasites and scar tissue in the usual way; it also goes in and out of the hippopotamus' mouth to remove any debris from between the teeth. This hippopotamus is strictly herbivorous, but the fish must run some risk of being accidentally swallowed by the hippopotamus as its mouth is huge and insensitive. *L. velifer* is the only recorded freshwater fish to act as a cleaner; all the other cleaner fish are marine. The Pygmy Hippopotamus (*Choeropsis liberiensis*) may also receive this service, but there have been few records of this. The pygmy is much rarer and shyer than its larger cousin and this fact could explain the lack of records.

The two species, hippopotamus and crocodile, have arrived at a solution to the same problem in very different ways. This is largely due to the proportion of time spent by each either in the water or on the river bank. The crocodile spends large parts of the daylight hours on the bank and is therefore accessible to the plover, whereas the hippopotamus spends the same daylight hours in the water and can be cleaned by a fish. Both species feed mostly at night and would not wish to be bothered by cleaners while they hunt or graze. The arrangement seems both simple and effective.

Birds and Ants
Several species of bird, for example jays, starlings and jackdaws, use ants to rid them of the troublesome parasites that are so abundant on all birds. This process has been recognised for a long time and is referred to as 'anting'. The bird picks up an ant very gently and pushes it into its feathers. The ant sprays formic acid when annoyed, and this kills the parasites. The ant can then be removed and replaced by the bird on another itchy area of its body.

Cow Birds and Cattle

The Brown-headed Cow Bird (*Molothrus ater*) is an example of a species that is at the same time a symbiont and a parasite. It is a North American bird which used to follow the enormous herds of bison which once roamed the continent. As man encroached on the bison's habitat, its numbers were decimated, eventually becoming close to extinction in the wild. When this happened the cow bird turned its attention to the domestic cattle that replaced the bison.

Cow birds feed in the same way as the African oxpecker and cattle egret already mentioned. They follow the cattle, hopping around the animals' feet, and feeding on the insects that are disturbed as the animals move through the grass. They also clean insects and other parasites from the host's body and act as a sentinel, warning of approaching danger.

When the USA was populated by bison, they moved in huge numbers across the continent in annual migrations for food. In order to be able to follow the bison's migrations, the cow bird had to establish a method of reproduction that did not involve it in

Another bird which performs a cleaning service for cattle and other herbivores is the brown-headed cow bird.

remaining in one place to incubate its eggs. To do this it turned parasite, using the cuckoo's method of laying its eggs in other birds' nests. This meant that the cow bird only had to remain in one place long enough to lay its eggs. The range of birds that the cow bird parasitises in this way is considerable; they have been known to lay eggs in the nests of more than two hundred and fifty different species.

A hippopotamus being cleaned by a small carp-like fish.

Now that the cow bird associates with relatively static domestic herds it is thought possible that it might change its breeding habits, building nests and rearing young in the normal way. In evolutionary terms the cow bird's change from a nomadic lifestyle to a more static one is only just beginning, but it will be interesting to see, in later years, whether the bird conforms to normal breeding patterns. Another reason for supposing they might change their habits is that the parasitic method they use at present is not an economic way of producing young. The cow bird has to lay many more eggs than other birds to maintain the species. Up to forty eggs are laid in various nests by individual hen birds, and this must prove quite a physical strain.

58

Cow Birds and Oropendolas

This type of symbiotic/parasitic association is taken one stage further in the relationship between the Giant Cow Bird (*Scaphidura oryzivora*) and the Chestnut-headed Oropendola (*Psarocolius cassini*). The oropendola and cow bird are both inhabitants of tropical, central America and are most commonly found in and around Panama. The cow bird family are nest parasites, as we have seen with the brown-headed cow bird; one of the host species in whose nest they choose to lay their eggs is the oropendola. This association is not a simple case of parasitism, however, as in some instances the presence of cow bird chicks is actually beneficial to the oropendola nestlings. This is most unusual, as normally the young of the host species are in danger from the parasitic chicks, either in competition for food, or because the parasitic young actually remove the host chicks from the nest.

Most nest parasites develop the ability to lay eggs that closely resemble those of their host so that the eggs are not rejected or destroyed. In 1968 scientists working at the Smithsonian Tropical Research station in Panama found from studying shells on the ground by nesting colonies of oropendolas that in some colonies (of 10–100 nests) the cow bird eggs were good imitations whereas others bore no resemblance at all to the oropendola eggs. Experiments were carried out to try to explain this. Different eggs and other objects were put into the nests to see if the birds were able to discriminate between true and false eggs. This was a considerable feat as the oropendolas' nests are almost inaccessible, 7–20 m up in the trees, hanging over the smallest branches. They are made of vines and grasses, hanging downwards forming a pendulous net. Ladders, long pincers and other tools were specially made so that the experimenters could remove and replace the nests intact.

By this method they were able to establish that some oropendolas were able to distinguish the cow bird eggs and other objects from their own eggs and others were not. The latter would incubate any eggs or even unrelated objects. Further observations showed that in colonies of non-discriminatory birds young oropendolas were often infested with the larvae of a species of bot fly and that this infection frequently caused the death of the nestlings. In colonies of discriminatory birds these parasites were very rare. It was also true that in non-discriminating colonies nests with cow bird young were less infested than those without cow bird chicks.

It was then observed that the cow bird chicks would snap at any small objects moving in the nest, including adult bot flies, thus

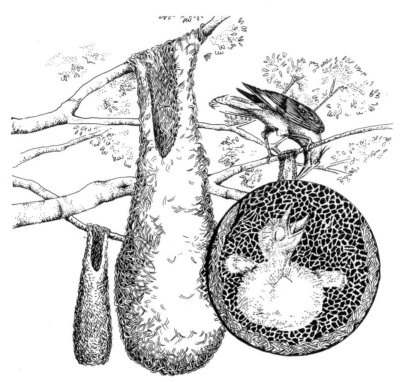

A giant cow bird chick in the pendulous nest of the chestnut-headed oropendola.

preventing the fly from laying her eggs. These chicks were also seen grooming the host nestlings and removing any larvae. The cow bird chick is well equipped to do this as it hatches a few days before the host young, and therefore has its eyes open and is more mature than the host chicks.

The next question was why the bot flies were found in non-discriminating colonies and not in the others. The locations of separate colonies of non-discriminating and discriminating oropendolas were studied. It was found that the discriminating colonies were established close to bee or wasp nests whereas the others were not. It is likely therefore that the presence of the bees or wasps inhibits the bot fly and encourages it to go elsewhere to lay its eggs.

From the evidence it would be fair to conclude that in certain environmental conditions the oropendola and cow bird are symbiotic, in that the oropendola gains a cleaning service in exchange for hatching and caring for the cow bird young.

Blowfly maggots
cleaning wounded
hide by consuming the
infected and dead
tissue.

Flies

The last cleaning symbiosis in this chapter is quite different from any of the previous examples. It concerns Bluebottles (*Calliphora erythrocephala*) and Blowflies (*Calliphora vomitoria*). Bluebottles and blowflies are the large-eyed flies that can often be found around dustbins and near animals. The female lays her eggs on decaying matter of all kinds (animal corpses, tainted meat or fish); anything that will provide food for the developing larvae. One of the places most favoured as an egg-laying site is a suppurating or festering wound. When the larvae hatch they feed only on necrotic tissue and pus, not on the healthy tissue surrounding the wound. They require the rich source of plasma and red blood cells found in necrotic tissue. The excretions of these larvae act in a similar way to a disinfectant and so also help to clean out the wound. By eating infected tissue and disinfecting the wound these grubs provide the host animal with a valuable service and could often save the host's life by preventing the infection spreading all over its body.

61

Blowflies were used in hospitals in the last century as wound cleaners. Infected wounds were sealed up with blowfly eggs and left for a few days, until the nurses could hear flies buzzing around. Then the bandages were removed and the infection usually gone. The American Blowfly (*Phoenicia sericata*) is especially efficient at this task and was a favoured hospital species. There are, however, several other species of blowfly whose larvae will eat any part of the animal and not only the necrotic matter. These flies are obviously of little benefit to the host and such a relationship could not be called symbiotic. This is a strange relationship as the host only has need of it when infected and normally has no control over the presence or absence of larvae. The flies are given a useful breeding ground for their larvae in exchange for a cleaning and sterilising service.

The examples of cleaning symbioses mentioned in this chapter are designed to demonstrate some idea of the depth and spread of this phenomenon. There are many more species that clean others, but most of them do not vary to any great extent from the ones described.

4 PROTECTION

The fight for existence in the animal kingdom has been described as the survival of the fittest, that is only those creatures that are healthy and well equipped to resist the dangers of life will be able to develop and reproduce. Survival is only possible when animals can feed and protect themselves. There are many symbiotic relationships in which the parties are better able to feed in the association than alone. There are also several cases where the motivation for joining forces with another species is that of protection rather than feeding. These examples are rather rarer than the feeding/cleaning kinds, but it must be remembered that there is often also a protective element in the cleaning associations. The instances quoted in this chapter are those where the outstanding motivation is protection.

Again, most of these examples are marine, but some concerning mammals and also a plant/insect association will be mentioned. The reason why so many symbiotic associations are concerned with the sea is a source of speculation. The marine environment is a very dangerous one for small creatures; only the very large and predatory fish are safe from natural predation. Small creatures are vulnerable all the time, day and night. They protect themselves as much as possible by sheltering in rock crevices or coral, by burrowing in sand or by living in enormous shoals. Nevertheless, even these forms of self-protection cannot eliminate the extreme danger of predation that they constantly live with.

Shrimps and Fish
The first example of protective symbiosis is quite extraordinary. One of the pair of species involved, the shrimp, seems so totally ill equipped for life in the sea that it is a wonder it survives at all. This continued survival must be largely due to its association with the other species. The Blind Shrimp (*Alpheus djiboutensis*) and the

The goby fish leads the blind shrimp so that it can feed in relative safety. The goby also shares the shrimp's burrow (below).

Gobid Fish (*Cryptocentrus lutheri*) live in the warm waters of the Pacific and Indian Oceans and the Red Sea. They inhabit areas of flat sandy sea bottoms where there is little or no coral or rock in which to seek shelter. The shrimp digs burrows in the sand to live in and shares the accommodation with a gobid fish. The fish uses the burrow as a temporary refuge during the day, and as a resting place at night. The shrimp is purblind and relies totally on the fish to guide it around. *A. djiboutensis* never leaves the burrow alone as it would be totally unable to sense danger. Instead it waits for the fish to swim close by, then it extends its antennae until it touches the fish's side. They both then leave the burrow in tandem fashion. The shrimp keeps its antennae touching the fish at all times, and in this way the two animals feed. At the first sign of danger the fish waves its tail from side to side and the shrimp picks up the vibrated message through its antennae. If the danger is imminent the fish swims into the burrow with the shrimp 'in tow'. In this way the two have developed a sophisticated communication system.

The benefit to both species is obvious. The fish has a bolt-hole for daytime and a safe resting place at night, both of which are vital in the kind of flat, sandy area where they live. The shrimp is allowed to feed in relative safety by the presence of the fish, and without this sharp-sighted assistance the shrimp's life expectancy would be severely limited.

Fish and Jellyfish
The next example is also marine, and concerns a tiny fish that takes great advantage in an association with a dangerous and unpalatable predator. The host is the Portuguese Man-o'war (*Physalia physalia*), a dramatic-looking siphonophore, or jellyfish. The man-o'war has very long tentacles streaming down from its mantle which can exist in such density that patches of the sea seem to turn purple. These tentacles have nematocysts (stinging cells) which produce a virulent poison. This is used to immobilise the siphonophore's prey so that it can be eaten at leisure. A small blue and silver striped fish called the Horse Mackerel (*Trachurus trachurus*) lives among the tentacles of the man-o'war. This little fish has never been observed in the open sea, but spends its whole life among the tentacles. It is not known whether the fish has developed a chemical immunity to the toxins from the nematocysts or it avoids touching these cells by constant movement. If it does the latter it must be in a continual state of alertness as the tentacles wave around in the water all the time.

The fish receives constant protection from the host, as the jellyfish is absolutely unpalatable to any other species. The man-o'war seems to use the fish as a brightly coloured lure to entice prey species. The siphonophore certainly makes no attempt to eat the horse mackerel, so it can be assumed that its presence is either desired or at least tolerated (*see* p. 66).

Crabs and Sea Anemones
There are many examples of different species of hermit crab using sea anemones as a way of camouflaging their shells. Only the most extreme example of this type of relationship will be detailed. The association between these two animals is so close that if they are separated for any reason they do not usually survive. The hermit crab is a crustacean just like other crabs, crayfish and lobsters, but it differs in that it does not have the hard exoskeleton that the others have as protection. The abdomen of the hermit crab is soft and twisted to the right to enable it to utilise an empty sea snail shell, which acts as a substitute skeleton. When they find an empty shell

they twist their bodies into it, leaving only head, pincers and two pairs of walking legs showing out of the shell's opening. Their bodies are soft enough to be crammed in this way but are rough textured which helps them grip the inside of the shell. This adhesion is so strong that it is impossible to extract the crab from its host without pulling it apart. As hermit crabs grow they normally have to move into larger shells, which is a dangerous business as while they move over they are unprotected and very vulnerable to predators.

Horse mackerel living within the tentacles of the Portuguese man-o'war.

This particular hermit crab (*Eupagurus prideauxi*) has developed a novel way of avoiding as many of these shell changes as possible. It uses an anemone called the Cloak Anemone (*Adamsia palliata*) as camouflage. The crab finds an anemone and fixes it onto its snail's shell. The anemone gradually grows until it has covered the shell, and then it extends from the opening to form a tube. This enlarged area allows the crab to grow more itself without having to change shells. Thus the dangerous shell changing times can be minimised. When they are forced to change shells, they often remove the anemone from the first shell and carefully place it on the new one.

The crab gains protection from the anemone in two ways. Firstly, the 'cloak' soon covers the host shell and makes it appear from above to be only a rather fast-moving anemone. This is excellent camouflage for the crab, as anemones are not a favoured food source for predators. Secondly, it is able to share in the benefits of the toxic tentacles which the anemone uses to stun its prey. Any predator that desires to eat the crab has first to run the gauntlet of the anemone's powerful sting. The anemone does not seem to benefit from the association in such an easily defined and concrete way. It does, however, share in the crab's food. The crab is a messy eater and small pieces of fish drift up to the anemone where they are easily eaten. The anemone also benefits from a constantly changing environment as the crab moves from place to place. Anemones are able to move on their own, but their progress is very slow. The crab on the other hand can scuttle about in the search for better feeding areas. This must improve the anemone's chances of a sufficient and varied diet.

Crabs and Sponges

A similar arrangement exists between the Sponge Crab (*Dromia*

sponge crab with
d sponge covering
s carapace.

vulgaris) and the Red Sponge (*Amoroucium*). It is somewhat different in that the sponge is a very primitive form of life, and also the crab has an amazingly strong drive to cover itself with the sponge. The sponge crab is a Mediterranean species living on the sea bed at depths of 10–30 m. When it finds a sponge it cuts off a section using its claws as scissors. It puts the piece of sponge onto its shell and holds it there by hooking a modified spike on the last two pairs of legs into the sponge. These sponges have the ability to regenerate or continue to grow even when severely damaged. The sponge soon covers the whole shell, moulding itself into the contours. As with the crab/anemone arrangement, this camouflage makes the crab appear like a walking sponge. Again, like the anemone, sponges are unpalatable as food.

The strength of motivation for the crab to cover its shell is surprising. They seem really to 'suffer' when deprived of a sponge. They have been seen to 'fight' each other when there is only one piece of sponge.

Experiments have been carried out in aquaria to test the strength of this motivation. Crabs were deprived of their sponges for a while and then offered cardboard, paper, cloth and so on as substitutes. The crabs approached all these substances in the same way, cutting them up and putting them on their shells. This indicates how strong the crab's urge is to cover its shell.

The crab receives similar protection to the hermit crab, already mentioned. The sponge benefits particularly as the crab is always on the move, giving the sponge a constantly changing habitat. This is more crucial to the sponge than to the anemone as the sponge is a filter feeder and therefore the change of water is highly desirable.

Tiny flagellate (beating) cells with cilia beat rhythmically inside the sponge's body and this action draws in water through the many pores. Food particles go in with the water and are retained inside the sponge for digestion. The filtered water is expelled through a large opening called the osculum. This is such an efficient method of feeding that it has been calculated that a sponge about the same size as a man's head will circulate 3500 litres of water a day. A sessile sponge must spend much of its time refiltering water that has already been inside its body, whereas a sponge on a crab's back will have 'new' water to filter all the time. It will also reap the benefits from the crab host's messy feeding habits, which provide an enriched environment for the sponge.

This association is then a slight variation on the crab/anemone example, and is included to give the reader an example of two different associations that have evolved along similar lines to solve a similar problem.

Antelopes and Baboons

The next example of mutual protection by symbiosis is a much more casual arrangement. It perhaps indicates the early stages of a symbiotic relationship which could, if other factors were right, advance into a closer more advanced association. The Impala (*Aepyceros melampus*) is a gregarious and graceful African antelope. It is distributed from the Cape to Uganda, living in large herds close to waterholes. It is one of the most vulnerable ungulates as it is a preferred food of lion, leopard, cheetah, hyena and hunting dogs. Baboons are a largely territorial species of monkey known for their aggressive behaviour. They are large and strong and when attacked have been observed to give leopards a fight. Troops of baboons of various species will often associate with herds of impala. Both species prefer to live near water, so they are naturally brought together. Their association is not purely accidental, however, as they actively seek one another out among all the other creatures around the water hole.

They gather together for a kind of teamwork protection system. The impala acts as lookout with its incredible alertness, excellent eyesight and hearing in addition to its well developed sense of smell. Once alerted, the baboon is able to protect both the impala and its own troop, as it is both powerful and fearsome. A group of baboons is quite able to scare away even the most powerful African predator, the lion. Both species can and do live without the other in some places, but in the dangerous environment of a watering area they are better able to survive if they can join forces.

Acacia Plants and Ants

The final example of protective symbiosis is quite different. It concerns a plant and an insect, and there is a kind of role reversal. The plant 'uses' the insect to protect it, just as many animals and insects use plant material to protect them. The Acacia (*Acacia sphaerocephala*) is a bush that grows in the African savannah. It has large vicious-looking black thorns 3 cm long, which give the plant some degree of protection from leaf-eating animals. The thorns are, however, hollow and therefore brittle; so any really determined animal can break off or flatten the thorns without too much difficulty.

Overleaf: Impala and baboons near a water hole. The impala with acute hearing and eyesight acts as lookout, and the aggressive baboon attempts to drive off any predators.

69

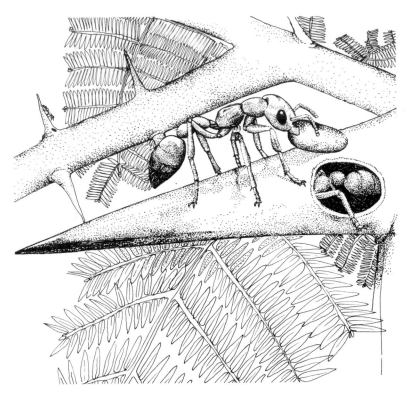

A species of ant (*Pseudomyrma*) has become adapted to live
in these hollow thorns and provides the plant with added protection. The ants have a virulent bite and when an animal attempts to
browse on the acacia the ants come out of their thorns and attack
the soft nose of the animal. This discourages most creatures with
the notable exception of the giraffe. Giraffes have long, strong,
prehensile tongues with which they strip the leaves off trees. Their
sensitive noses do not come into contact with the plant material
and therefore they are not badly bitten.

The ants nest in the thorns, going in and out through tiny holes
that they make at the base of the thorn. They feed on a diet of
sugar, oils and albumen provided by the plant. This food supply
means that the ants never have to leave the plant, and the thorns
make a well-armoured home for them. The substance that the ants
feed on is specially manufactured by the plant, which has no need
of it for its own purposes. It is intended to keep the ant colony and
deter it from straying, which means that the ants are always 'on
duty' to protect the plant. This is a most unusual example of a plant
enlisting the services of an animal, and it certainly is not accidental

as the plant produces this food for that specific and only purpose. This relationship provides both plant and ant with solid protection, with the added benefit to the ant of a constant food supply.

There are other examples of ants and acacia living together in similar ways, for instance *Pseudomyrmex ferruginerea* and *Acacia cornigera*, which are so dependent on one another that they are never found alone.

Protection is vital to all but the strongest, largest or most unpalatable in the plant and animal kingdoms. Some creatures are better fitted to survive because of their physical make-up or the particular environment in which they live. The examples in this chapter all concern creatures or plants that are vulnerable to predation of various kinds. The soft-bodied hermit crab, blind shrimp, tasty sponge crab, tiny horse mackerel, impala and acacia all live in dangerous circumstances which make them liable to be eaten if they do not take extra precautions. None of them are well equipped by size or strength to repel attackers, but they have all found added protection by association with other creatures. The fact that they continue to do this from generation to generation means that the arrangement is successful in prolonging their chances of survival.

5 POLLINATION & SEED DISTRIBUTION

Pollination is the method by which flowers reproduce sexually. It is the movement of pollen from the male part of the flower, where it is manufactured, to the female part where it fertilises the ovum. All flowering plants reproduce in this way, but they have evolved a multitude of different methods of transporting pollen. The male part of the flower is called the stamen and it produces the dust-like substance called pollen, which is analogous to mammalian sperm. The female organ is the carpel which contains the stigma, style and ovary. The stigma receives the pollen which then grows down the tube-shaped style, to the ovary, fusing with the 'egg' or ovum and thus producing the seed which becomes the next generation of plant.

There are three main types of pollination: self-pollination, wind pollination and pollination by animals. Self-pollination occurs when the pollen from a flower fertilises ova from that same flower. In these plants the stamen are usually above the stigma and the ripe pollen just falls onto the stigma. There are problems with this method, as no new genetic material is introduced, which is the main element in sexual reproduction. This means that if the environment changes, requiring alteration on the part of the species, there is no genetic reservoir enabling the species to respond. In normal reproduction the offspring's genetic make-up is produced by the fusion of both parents' genes, thus giving strong dominant characteristics the chance to be introduced from the vast variety of features available.

Some plants are pollinated by the action of wind blowing pollen from one plant to another. This can only happen successfully when several blossoms of the same species are in close proximity to one another. Only the pollen of the same species of flower can pollinate it, so all the other types of wind-borne pollen that fall upon the stigma are wasted. It is quite obvious that enormous quantities of

74

A humble bee polli-
nating a daffodil. The
bee takes nectar from
the base of the petals
and during the process
inadvertently collects
pollen which it trans-
fers to the next flower.

pollen need to be produced and a very great deal of it will be lost by falling to the ground and never reaching a receptive stigma.

The most common and effective form of pollination is the cross-fertilisation brought about by insects, birds and some bats. These animals brush into the stamens and inadvertently collect pollen on their bodies. As they move from flower to flower they distribute the pollen onto the stigmas of other plants of the same species. This reduces the quantity of pollen that needs to be produced and extends the distance that the plants can afford to be away from one another to the normal range of the animal pollinator.

To get the benefit from animal pollination the flower has to attract either a specific creature or any one that is able to transfer pollen. This has resulted in the evolution of the fantastic variety of colours, shapes and perfumes that are present among the flowering plants of the world. Some flowers are of general beauty and fragrance in order to attract any passing insect in order that the flowers may be pollinated. Others aim to attract only the one species of insect that is able to fertilise it. These last varieties of flowers can imitate either the pheromone (sexual perfume) of the insect or its appearance. Some plants, such as the bee orchid, are able to mimic a specific insect so well that males of that species persistently attempt to mate with the flower, believing it to be a female. In other cases flowers are marked in such a way as visually to guide the insect into the pollen-bearing stamens. These marks are often called honey guides. Most plants produce nectar as another inducement to insects and birds; this is a sweet, sticky substance found at the base of the petals which is a most attractive food.

In instances where the plant needs the services of only one species it is totally reliant on that one animal being present and willing to transfer pollen. If for any reason the essential insect moves from the area or is made extinct the plant will soon die out. Some examples of this extreme symbiosis are detailed in this chapter as are some less specific relationships. There is every degree of symbiosis from the occasional pollination by insects that normally do not pollinate to the total reliance already mentioned.

It is important for the flower that it receives pollen from another of its own species and not the pollen of a different species. It has been observed that insects favour a specific flower while that species is in blossom, usually only visiting that type and no other. This is called 'bee constancy', and it is also common to birds and other pollinating animals. They do not visit one species of flower at one time for altruistic reasons, but rather because it is far more efficient for them to concentrate on one plant and learn the way to remove nectar with the greatest speed and efficiency. Some mechanism must control this behaviour within the insect, but it is not yet understood how the system works. It is, of course, to the great benefit of the plant that the pollinators should behave in this way, as very little pollen is wasted and the maximum number of blossoms are fertilised.

The colours of flowers are not arbitrary but constitute another way of attracting insects. The colour vision of insects is often rather limited, that is they only 'see' one area of the spectrum. By being coloured to attract one type of animal, the plant can select the most efficient pollinator; for example bees are thought to be attracted by shades of blue whereas butterflies prefer red. Some plants mimic odours that are attractive to various insects. This can either be a perfume designed to copy the scent of a sexually receptive female or a smell that will attract an insect that does not normally act as a pollinator. For example, the apparent smell of rotting meat will attract several species of fly, and in crawling around the flower looking for the source of the smell the fly will pick up and transfer pollen.

It is not only insects that have the ability to pollinate, although they far outnumber the few birds and bats that do so.

As well as assisting plants with pollination, many species of animal and bird help in the later process of seed distribution. It is important that the seeds of a plant do not just fall to the ground around the adult. They would grow much too close to each other and therefore would not be able to develop into strong, healthy plants. Some means of transporting the ripe seeds to a different location is needed. Some seeds are aerodynamic so that they will blow away in the wind and be scattered. Some will float when they fall on water and can be transported downstream until they land and grow in the banks of the river. Some seeds have spines and bristles which enable them to stick onto the bodies of animals and birds that brush into the plant. These sticky seeds irritate the host and are groomed out of fur or feather and thus fall to the ground some distance from the parent. None of these methods are truly symbiotic but the seeds that are surrounded by edible fruits are often transported by animals in a symbiotic relationship. The plant gains the scattering of its seed and the animal receives a valuable food. It is essential that the seed inside the fruit is strong enough to withstand the normal digestive processes of the animal and can pass through it without being destroyed. There are so many examples of this behaviour that it is not possible to list them all. Virtually all fruit-eating creatures regularly help transport these seeds. One example of a rather extraordinary and extreme seed transportation, which is described more fully later in this chapter, involves the deliberate gathering and 'planting' of the seeds of a pineapple by ants.

The consequences of both pollination and seed distribution performed symbiotically by various creatures is of considerable value to the world as a whole. Sexual reproduction by cross-fertilisation is by far the best method of reproduction to allow evolutionary responses in plants. None of the other methods involve the introduction of new genetic material, and as a result the genetic reservoir of the species is weakened. It is not possible to over-estimate how important this cross-fertilisation is to the whole flora. There would not be the variety or splendour of vegetation that we see today if cross-fertilisation did not occur. Variety cannot be induced in a plant that has only one set of genetic information to pass on to its offspring. It is necessary for two different sets of

Overleaf: A butterfly performing the same pollinating function the bee shown on p. 75.

77

genes to come together to produce the slight changes from generation to generation that gradually evolve to create new strains.

Yucca Plants and Moths

Probably the most extreme case of interdependence between an insect and a plant is between the Yucca plant (*Yucca* spp.) and a small moth, the Yucca Moth (*Tegeticula yuccasella*). They are so reliant on one another that neither species could survive without the other. They inhabit Mexico and adjacent Central American countries. The yucca has lush, fleshy leaves that grow in a crown, and long stems with waxy white blossoms. Each stem carries many bell-shaped flowers which droop downwards. The yucca cannot self-pollinate and the bell shape of the petals means that airborne pollination is most unlikely. The plant therefore relies totally on the small yucca moth to pollinate it. The moth does not transfer the pollen as a by-product of feeding, as many insects do, but actively collects and delivers the grains from stamen to stigma.

The female moth carries out the pollination, collecting the pollen with a specially adapted proboscis looking like a pair of curved tentacles. The yucca pollen is slightly sticky and she is able to work it into a ball two or three times the size of her own head. She balances this ball of pollen on her front legs and flies off to another flower. She is most careful not to 'spill' pollen onto the stigma of the first plant, and the tidy ball she is carrying enables her to be very precise about which blossoms she fertilises. In this way she ensures that the plants are cross-fertilised.

The female moth is now ready to lay her eggs. When she arrives with the pollen at the second flower, she first lays some of her eggs in the flower's ovary. Then she rubs the ball of pollen onto the stigma, ensuring pollination. She then flies onto another flower and repeats the process until she has laid all her eggs and delivered the pollen (*see* p. 82).

Pollination begins the seed-making process, and at the same time as the seeds are forming the moth larvae are hatching. After hatching the larvae feed on the seeds which are inside the seed pod with them (formed from the ovary). The larvae only eat about twenty of the two hundred or so seeds before they grow too large, bite their way out, spin a thread and lower themselves on it to the ground. They then dig their way into the soil, spin a cocoon and only emerge when spring comes. It has been found that the yucca moth larvae are totally dependent on these particular seeds, and they have never been successfully reared on any other diet.

This relationship is very finely balanced between success and the extinction of both species: if either species is unable to join in the partnership no further generations of either plant or moth would be produced.

There are many mysteries concerning this unique relationship that are left to be explained. For example how does the moth know that she should take pollen from one plant to another? What mechanism exists to restrict the number of eggs that the moth lays to that which the individual blossom can safely maintain? Finally, there is the accuracy with which these events are timed. The flowers are only open for a very short time and the moth must be at point of lay at that exact time. If she is not ready the pollen would be wasted and no eggs laid. As well as that timing accuracy the other element that involves extreme coincidence is the development of the eggs to larvae and the fertilised ova to seeds. They both have to be at the correct stage of development at the same time, or the process would fail and no young of either species would be produced.

The behaviour of the moth in this association appears to be considered and premeditated, but this kind of anthropomorphic interpretation cannot really be justified. The action of the moth is a demonstration of natural selection in action; those creatures most able to survive and reproduce are those that do so successfully. The relationship between the plant and moth is economic of effort, eliminates waste and enables both species to survive and reproduce. The moths that help their essential food host to survive are the most likely individuals to have young, and those new moths will carry the instinctive desire to pollinate the yucca plant. This has to be an instinctive response as the female moth has no opportunity to learn by example. Any moth that does not pollinate will fail to provide her young with the food source they need and therefore they will die; in this way the non-pollinating characteristic will die out with her young.

Apart from the extreme nature of the association between the yucca and its moth it is also unusual in that the moth does not visit the flower for nectar or any other foodstuff. Most insects' main intention when going to a flower is to get at the nectar hidden at the base of the petals. The yucca moths' intentions are more subtle and complicated than the simple instinct to feed.

Bees and Orchids
An example of a species that is 'tricked' into pollinating a flower is

the Euglossine Bee (*Euglossa meriana*) which is duped by the orchid *Stanhopea grandiflora*. This orchid does not produce nectar, so food is not the attraction for the bee. It seems intent upon collecting scent from the flower. Only male bees visit the orchid, and they collect the scent with specially modified forelegs, storing it in a pocket on the hind legs. It has been suggested that the male bee uses this perfume to attract members of the opposite sex. In the course of collecting the scent the bee transfers pollen from one flower to another. It is presumably the perfume that the flower emits that attracts the bee in the first place. There are several different kinds of stanhopea orchid, each of which gives off a different scent and consequently attracts a different species of bee.

There are other examples of species that are tricked into pollination of flowers, but most of them are not symbiotic as the animal usually does not benefit from the arrangement. There are insects that are conned into thinking that a blossom is a female of their own species, and in the act of 'mating' with the flower they collect and transfer pollen. Arum lilies have the ability to trap insects

A yucca moth which has just delivered a ball of pollen into the female part of the yucca flower.

broad-billed
umming bird about
o take nectar and
onsequently pollinate
tube-like blossom.

inside the cone of their petals for as long as it takes the insects to pollinate them. In both these and other instances only the plant benefits from this association.

Humming Birds

Humming birds (Trochilidae) visit flowers to collect nectar which is their only foodstuff. There are over three hundred known species of these tiny, fast-moving birds that inhabit most of the new world. The largest is the Giant Humming Bird (*Patagona gigas*) of the Andes, which is 20 cm long (half of which is its tail). The smallest is the Bee Humming Bird (*Mellisuga helenae*) from Cuba, which is only 5 cm long. Humming birds get their common name from the sound of their wings which hum as they hover over flowers to drink the nectar.

Humming birds do not settle to take their food, but feed while on the wing. They have developed the ability to hover in the air, and even to fly backwards. They have long curved bills which allow them to drink the nectar which is usually at the base of the flower petals. While their bills are taking in the nectar the bird's head is pushed deep into the flower where it collects pollen. As the bird moves on to the next flower the pollen is transferred to the stigma. Humming birds consume vast quantities of nectar, as the continuous rapid beating of their wings uses up a great deal of food energy, and so they visit many flowers. They seldom land, and feed throughout the day. This makes them of considerable value to the plants as prodigious pollinators.

Ants and Pineapple Plants

Seed distribution has already been mentioned in general terms in this chapter. An example of animals helping in the scattering of seeds occurs in a species of ant which collects and 'plants' pineapple seeds. Several species of the family Bromeliaceae live in the flood basin of the Amazon in South America. In this region water levels rise regularly by several metres, thus drowning large numbers of plants and terrestrial animals. These floods would certainly destroy ant nests, and so the local ants have developed a novel method of avoiding the destructive effects of the water.

The ants have moved up into the trees and live in nests which

Ants living in the tropical gardens of pineapple which they have grown from seed.

they construct for themselves. They search the ground for seeds of the pineapple, take them up into the trees and plant them in mud that they have already transported. These seeds germinate and grow, attaching themselves to branches around them and forming a strong nest-like network of roots, stems and leaves. This plant provides the ants with a much stronger nest than they could make from the existing plant material, and also a tropical garden in which to feed and shelter. This new home means that they no longer have to risk the dangers of the water and can live their lives above flood water levels. The bromeliad benefits by the ants taking it up into the trees as it is no longer in danger of being washed away. It is also much nearer to the sunlight that it needs in order to photosynthesise. The ground level in these tropical forests can be very dark as the trees grow so close together that little light can filter down.

Both species gain considerably from this association, which is simple but highly effective. The ants must have arrived at this process of collection and planting bromeliad seeds by some means of trial and error, but by now it has been established as a successful venture.

All these instances of pollination and seed distribution, as well as the hundreds that have not been detailed, are of great importance. An effective method of cross-pollination is vital for the variety and vigour of most plant species, and this is in the main brought about by animals of various kinds. These creatures are directly responsible for the amazing variety of flowering plants that exist today. Of similar, if not equal, importance, is the process of seed distribution that is carried out by so many animals. Without this scattering, plants would grow too close together and suffer the consequences of overcrowding.

6 ALGAE

Photosynthesis is the process by which all green plants obtain their nourishment. As the name implies, light is required in order for this to take place. Photosynthesis is common to plants of all habitats—fresh, sea water and terrestrial species. During the process the plant converts the carbon dioxide in the atmosphere into carbohydrates. The need for light to enable a plant to photosynthesise explains why they do not live at great depths in the sea and why they proliferate in clear, fast-flowing rivers.

When we refer to plants as photosynthesisers it must be remembered that some tiny, often unicellular organisms are technically plants. It is their ability to photosynthesise that enables taxonomists to categorise them as plants, as no animal can feed by this means. This chapter is not concerned with the larger plant species, but rather with the relationships between various animals and fungi with tiny plants. In these associations, without exception, the host species uses the plant's ability to photosynthesise as a way of obtaining extra nutrients. There are some essential physical characteristics that the host must have in order to enable the plant to photosynthesise. The host must have a translucent body; or at least the part in which the plant lives must have enough clarity to allow rays of sunlight to pass through. The host habitat must also allow it fairly constant hours of sunlight.

All the examples of symbiosis in this chapter concern algae of various species. Algae are plants and therefore do photosynthesise. They are very simple forms of life, often extremely adaptable to different environments.

Hydras

The first example concerns the Green Hydra (*Hydra viridis*) and a form of alga called *Chlorella*. The hydra is a small freshwater animal, long and tube-like with one end closed, clinging onto

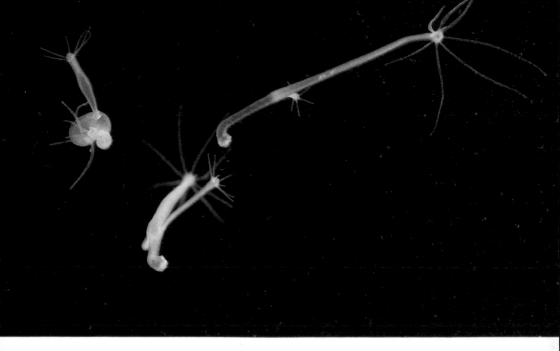

Three green hydra, the colour of which indicates the presence of algae. Non-symbiotic algae are translucent and colourless.

plants or rocks. The other, open, end is surrounded by waving tentacles, which move through the water with their tiny stinging cells (nematocysts) ready to paralyse prey, usually small crustaceans. The tentacles then gather together the stunned bodies and bundle them into the 'mouth' opening. Hydras have a body wall two cell layers thick, the ectoderm (outer layer) and the endoderm (inner layer) separated by a non-cellular layer called the mesoglea. Food is taken in and waste excreted through the same opening.

Hydra viridis associates with unicellular algae called chlorellae which, because of their green colouration, give this hydra its name. The algae are taken into the host's body via the opening along with the animal's normal food and are absorbed into the endoderm. The algae is able to photosynthesise from this position as the outer layer of the hydra is translucent. The hydra is able to absorb the oxygen and carbohydrates that the photosynthetic process produces which are surplus to the requirements of the algae. This can form a substantial part of the hydra's food intake and is not dependent on suitable prey being present. The hydra is able to move around to give the chlorellae the best chance of utilising as much sunlight as there is, and therefore extend the period in which the algae can photosynthesise.

Unusually, *Hydra viridis* is able to pass on to the next generation a full complement of chlorellae and this gives the offspring the same advantage enjoyed by adult green hydras. The hydra's reproduction is called 'budding'; the adult pushes out a tiny tube-like protuberance from the long side of its body, and as the tube grows it develops and when fully formed breaks away from the 'parent'. The offspring of *H. viridis* have the usual chlorellae in the endoderm and already photosynthesise.

To establish how beneficial the algae are to their host, experiments have been conducted comparing the ability of two kinds of hydra to withstand hardship. The second species of hydra does not form a symbiotic relationship with algae, but apart from this fact they are remarkably alike. When food was plentiful the growth rates of both kinds of hydra were seen to be similar. However, when the supply of nutrients available to the animals was reduced, the symbiotic hydra (*H. viridis*) grew considerably quicker than the non-symbiotic one. When its environment was completely starved of nutrients (but normal daylight continued), *H. viridis* lived much longer than the albino (non-symbiotic) variety. This seems to substantiate the fact that the hydra gains appreciable quantities of nutrient from the algae.

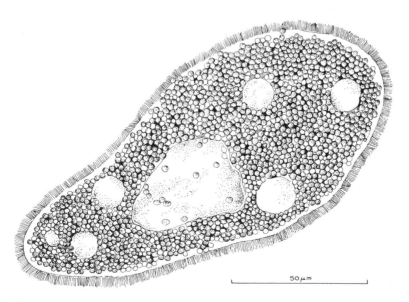

Paramecium bursaria with its symbiotic chlorellae.

Protozoa

This kind of relationship can be formed with even simpler, more primitive forms of life. The protozoan *Paramecium bursaria* is a freshwater ciliate that looks a little like an amoeba, but with a stiff outer layer giving it distinct hind and fore ends. It is a very simple animal that feeds by spreading itself around food particles, digesting them and moving away, leaving any waste behind.

Paramecium bursaria takes in algae (*Chlorella*) via the food vacuoles (similar to air bubbles) and in some way, as yet unknown, the algae escape into the cytoplasm of the animal. The chlorellae gravitate to the periphery of the cell from where they can reach relatively unfiltered sunlight. It is not understood how the protozoan is able to differentiate this chlorella from other similar organisms and thereby avoid digesting it. It seems extraordinary that such a simple organism is able to make such subtle choices.

The algae are able to carry out photosynthesis and provide their host with the carbohydrates and oxygen that are surplus to their needs. Experiments similar to those mentioned for *Hydra viridis* were carried out on symbiotic and non-symbiotic *P. bursaria*. The results were similar, in that the symbiotic ones were able to live longer in impoverished circumstances than the non-symbiotic ones. Even a small, primitive, simple animal such as this protozoan is able to utilise the services of a plant and gain valuable nourishment from it in exchange for providing a constant environment.

Flatworms

The flatworm *Convoluta roscoffensis* lives on beaches off Brittany, France, and the Channel Islands. It lives between high and low tide marks in rock pools or damp rock crevices. At high tide it burrows down among the grains of damp sand to avoid desiccation. When newly hatched the tiny worm ingests plant material in the normal way, and at this stage it is quite colourless. It allows the useful algae to separate out from the normal food and invade its body. When enough of these algae have done this, the worm changes from colourless to a slimy green.

The minute algae are able to photosynthesise from inside the host, as its ectoderm is clear. When a full complement of algae has been absorbed, the flatworm stops feeding and relies totally on the algae for its food supply. As a result of this behaviour the worm resembles a plant almost as much as an animal, as it must provide the plant with as much daylight as possible so that it can photosynthesise. This relationship has moved on several stages from

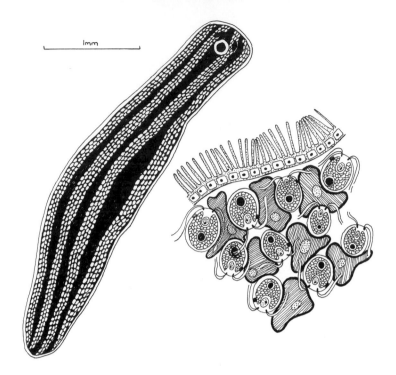

Convoluta roscoffensis.
The enlarged area on
the right shows the
symbiotic algae.

Hydra viridis and *Paramecium bursaria*, in that the algae also convert the worm's waste products, mainly uric acid, into re-usable food. Thus the nitrogenous compounds are continually recycled. The worm also digests any surplus or dead algae and thereby keeps the population under control and healthy.

The algae are provided with a moving home which enables them to reap maximum benefit from sunlight. They also receive a certain amount of protection from severe weather and a good supply of nitrogen, phosphorous and carbon dioxide from the worm's metabolic waste. This relationship is similar to the two previous examples, but shows how a relationship can develop towards greater efficiency and dependence. The flatworm actually stops feeding and relies on the algae for all its nourishment as well as depending on the algae's ability to recycle its waste products. These two facts increase the sophistication of the relationship and also the worm's dependence on its tiny plant invader.

Sea Slugs and Seaweed
A similar relationship exists between the sea slug (*Elysia viridis*) and chloroplasts from the seaweed *Codium fragile*. There is an interest-

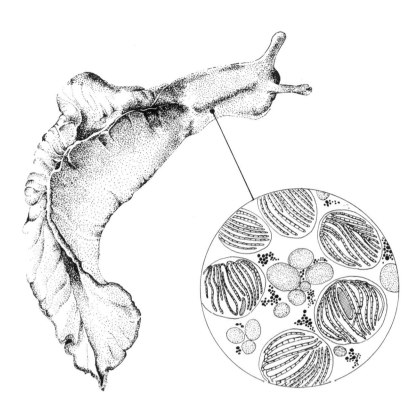

lysia viridis. The detail
on the right shows the
chloroplasts.

ing difference between *E. viridis* and *Convoluta roscoffensis* in that
the former actually seeks out its plant partner and does not simply
rely on the possibility of ingesting such organisms.

E. viridis is a small marine slug that is colourless until it has
ingested enough plant material to turn it green. The slug does not
utilise algae as in previous examples but seeks out chloroplasts
(from within chlorophyll-bearing cells) from the seaweed. The
chloroplast is an organelle, that is a specialised part of a cell acting
as an organ. The slug is able to puncture the cell of the sipho-
naceous seaweed *Codium fragile* and suck out its contents. These are
taken into the digestive system of the slug and in some way
separated out, invading the animal's body. Once again it is not
understood how this is done, but the chloroplasts are particularly
robust which might help them remain intact. The chloroplast is
able to continue life inside *E. viridis* and photosynthesise in the
usual way. This in itself is quite extraordinary as one tiny part of a
cell is totally self-contained and not reliant on the rest of its cell for
life. It was only fairly recently discovered that this slug did not
contain algae but the chloroplasts of a seaweed, and many scientists

thought it most improbable that the chloroplasts could continue life in this way.

The series of events that involves the active collection of chloroplasts, the fact that they remain intact after the digestion processes have operated on them, their ability to pass into the slug's body, and finally to live and photosynthesise there, is quite staggering, especially as it all happens in so humble a creature.

Lichen

The last example in this chapter of a relationship with algae is quite different. It does not concern an animal but rather associations between algae and fungi. Lichen is the name given to the product of an algal and fungal formation. The algae invade the fungi and the two together are called a lichen. This is an example of a third 'species' being formed by the joining together of two symbiotic partners. Over 18,000 different species of lichen have been described, found in most regions of the world. Most lichens are crustose, forming a thin, flat crust, but some are foliose or fruticose, shrub-like in appearance. These plants generally live in inhospitable habitats, such as on walls, gravestones and trees. Because of this they have to be resistant to extremes of temperature and rainfall as well as being able to exist in barren, low nutrient conditions. Lichens are very slow growing (1 mm per year is usual), but can live for long periods. Lichens of 200 years old are common and there is some evidence that Arctic lichens may be up to 4500 years old.

Although they do not have the ability to conserve water they are not destroyed by lack of it. This is necessary because the sheer vertical surfaces on which they live are often dry for long periods of time. When dry their metabolism slows right down, and they are then dormant until they re-hydrate with rainfall. Dry lichens are also very resistant to heat and can exist on rocks with a temperature of 50°C or more. These facts give some indication of how different lichens are from other 'plants'. They do not have roots or need soil for their survival, and are not susceptible to drought or extremes of temperature.

Lichens are the product of the association of one particular alga with a specific fungus; the algae are all either green or blue/green and unicellular or with simple short filaments. Normally the algae are restricted to a thin layer just under the surface of the fungal hyphae, but with a few species the algae and fungal filaments join throughout the thallus. As they live so close to the surface the algae

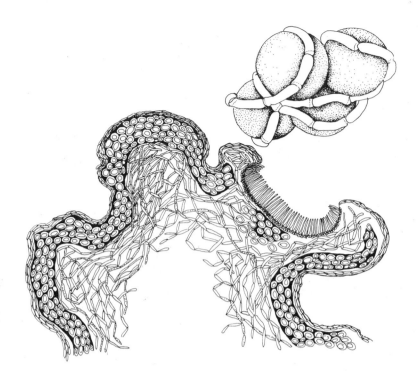

typical lichen
showing the algae and
the fungal hyphae.
The magnified section
(top right) shows how
the fungal threads
wrap themselves
round the algal cells.

are able to photosynthesise and provide the fungi with carbohydrates in the normal way. In this way the lichen gains its nutrients
even though the algae makes up less than 10 per cent of the whole
mass. There is no evidence that compounds move in the other
direction (from fungi to algae), but it is believed that the fungi
provide protection and enable the algae to colonise a much wider
range of habitats.

During the formation of a lichen both the algae and the fungi
undergo physical changes. The fungi are thought to cause the
outer cell layer of the algae to become permeable which allows the
transport of nutrients through this layer to the fungi. The existence
of so many lichens indicates that many algae/fungi symbioses have
arisen during the process of evolution. Some of the algae (for
example *Trebouxia*) that go to make up lichens are only to be
found within a lichen partnership, although they were presumably
once free-living. Lichens are very good indicators of pollution and
are used by ecologists for this purpose. They store minerals, sugars
and accumulate radio-active fall-out, and so can be examined and
'read' for any pollutants in the atmosphere.

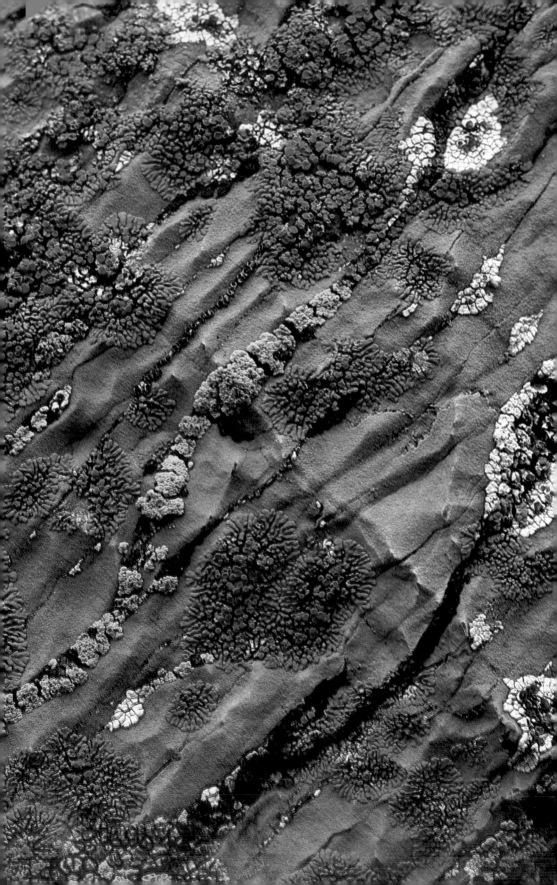

everal species of
chen showing the
arsh environment
n which they live.

The formation of a third species from two others is, as already stated, as far as it is known, unique in nature. The lichen has by this association been able to colonise the most unlikely and inhospitable habitats. Lichens give weight to the adage that nature abhors a vacuum; plants and animals seem determined to take advantage of any environment or circumstance that can possibly sustain life. Lichens help in the continuous process of rock erosion, as lithophytic (rock-inhibiting) lichens destroy the rocks on which they live. This is an extremely slow process, but nevertheless extraordinary, as most species are concerned with the well-being of their habitat rather than being an active agent in its destruction.

Although they are of limited commercial value these days, lichens were once of great importance to the populations of the northern regions. Even now several species of lichen (often called 'reindeer moss') are used as fodder and bedding for cattle, and for pharmaceutical and dyeing purposes. Three or more commercial antibiotics are prepared from the usnic acid found in lichens.

The kind of symbiosis displayed by lichens has sometimes been thought of as retrogressive— as a species' last resort. It is held that only the weak species that cannot fend for themselves resort to these associations. This is one viewpoint, but a counter theory suggests that only if symbioses are formed can such inhospitable habitats be colonised.

This chapter has been concerned with algae and chloroplasts providing fungal and animal hosts with nutrients by photosynthesis. It could not be said that the examples quoted are of as much importance to the rest of the flora and fauna of this planet as other symbiotic relationships. If any or all of these animals and lichens were destroyed the effects would hardly be cataclysmic; but each one has its place in nature and forms a part of a cycle of life. The lichen particularly is important, as mentioned before, as an accurate measure of pollution and as food for some Arctic creatures, especially reindeer.

There are many aspects of these symbioses that are not yet understood, but more study is at present in progress and it is hoped that we shall soon know how these simple creatures are able to differentiate between algal associates and almost identical algal foodstuffs.

7 MICRO-ORGANISMS

Bacteria have for many years received something of a bad press. For most people the very word sums up a picture of an unseen horde of uncontrollable, unpleasant invaders. So many diseases are caused by various bacteria that it is hard to see them in any other light. This is somewhat out of proportion, though; there are so many bacteria in the air we breathe, the water we drink and the food we eat that if they were all harmful we would not be able to live for more than a few moments. Most bacteria are either benign or positively beneficial, and it is only a tiny minority that cause any real trouble. The life of modern man would be considerably impoverished if bacteria did not exist, and indeed might not even be possible.

One of the strangest elements in the story of symbiosis is the extraordinary fact that none of the herbivores of this world can digest the grasses that are their normal diet without the aid of intestinal bacteria. The consequences of this are considerable; without bacteria these animals could not exist. That would remove from the face of the earth all the domestic sheep, cattle, goats and horses as well as the wild herbivores such as antelopes, gazelles, zebras, elephants and rhinoceroses. In addition to these, several kinds of rodent, marsupial and monkey could not exist. The majority of carnivores would therefore have no food source and the bulk of man's dietary protein would not exist. The effects on plant ecology and evolution would also be cataclysmic. In short, the world would be an entirely different place. The plant material on which these animals feed is composed largely of cellulose—the tough cell walls of the plants. Herbivores do not have the enzymes to break down cellulose but rely on the bacteria and protozoans in their guts to do it for them. They can then absorb the breakdown products from the plants (proteins and carbohydrates) as well as the bodies of any dead micro-organisms. These symbiotic micro-organisms live in various parts of the gut, according to the type of herbivore, but they perform essentially the same function for all.

Herbivores

Domestic cattle provide a typical example of this symbiosis. In cows, the stomach is divided into four chambers. The grass that the cow eats passes into the first chamber or rumen, where it is fermented by the micro-organisms present. This fermentation produces large quantities of methane gas (500 litres of gas in one day for each cow) and so conditions in the rumen are anaerobic (without oxygen). This anaerobic environment is essential for the survival of the dense population of bacteria and protozoans which live in the rumen. The relationship does not end at breaking down plants for digestion by the host, however. The bacteria have a very short life span of about 20 hours and when they die the cow can digest their protein-rich bodies. As the cow's diet of grass is low in protein, this extra nourishment is vital and can provide the cow with 20 per cent of its protein intake. This is therefore a very complete relationship. The cow cannot survive without the bacteria and the bacteria need this type of anaerobic environment with constant supplies of cellulose. The micro-organisms are so adapted to this anaerobic environment that they die in the presence

A normal population of micro-organisms in the rumen of a cow.

97

of oxygen. Both parties are thus ideally suited by the arrangement.

Other herbivores have differently designed alimentary canals and their micro-organisms may be housed in different parts of the system, but basically they perform the same vital function for similar rewards. Other animals, for example nursing piglets, require micro-organisms to provide them with an enzyme that breaks down their mother's milk sugar from lactose to glucose. Without this they rapidly get hypoglycemia and decline instead of grow.

It is not only herbivores and piglets that have intestinal micro-organisms; virtually every living animal hosts millions of such creatures. Most of these are not only harmless to the host, but the host may suffer loss of health if they are killed off. It would not be in their interests to harm the host because by doing so they would threaten their own existence. For the same reason, even in parasitic organisms there are usually self-limiting mechanisms which ensure that they do not kill the host.

Blood-suckers
Many blood-sucking animals are dependent on gut bacteria for digestion of the blood that they feed on. For example the Medicinal Leech (*Hirudo*), much favoured of Victorian doctors as a cure-all for their unfortunate patients, uses micro-organisms in this way. *Pseudomonas hirudinis* is the micro-organism involved and it takes this relationship one stage further by producing an antibiotic which stops any other micro-organisms growing in the leech's gut. Some other blood-suckers need gut bacteria to synthesise vitamins for them. Mammalian blood is low in vitamins and without those synthesised by their bacteria the host blood-sucker would not be able to grow and develop. Lice (*Pediculus corporis*) and Bedbugs (*Cimex lectularius*) both use bacteria in this way.

The relationship between these parasites and their bacteria is similar to that between herbivores and their micro-organisms.

Pogonophorans
One of the most unlikely creatures to exist in the world is also host to symbiotic bacteria. There is a phylum of the Protochordates (very distant ancestors of vertebrates), called the Pogonophora, whose members live close to the warm water vents in the earth's crust, at the bottom of the sea. They can be up to 1.5 m long. They live in tubes of their own secretion. The reason that they are so extraordinary is that they have no mouth, no intestine and no

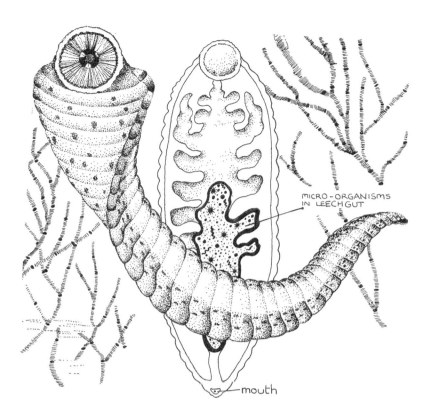

MICRO-ORGANISMS
IN LEECH GUT

mouth

A leech. The cut-away section shows characteristic micro-organisms in the gut area.

anus. They do have nerves and blood vessels, embedded in dense tissue which houses large numbers of bacteria. It seems incredible that an animal with no visible means of feeding or excretion can survive at all, especially as the hydro-thermal vents where they live are very low in organic material.

Some recent work has been done on this problem by three researchers from the Scripps Institute of Oceanography in California. They have decided that as the pogonophorans seem to have no external sources of nutrition they must have evolved an internal one. These scientists believe that the bacteria in the animals use some chemical source of energy in the water from which they are able to produce organic compounds which they then pass on to the host. It is possible that the bacteria have enzymes which allow them to make organic compounds by oxidising hydrogen sulphide and binding carbon dioxide. The precise method and how the organic substances are passed on to the host remain to be seen. This is very new research and by no means all of the questions have been answered yet. It is speculated that this bacterium might be closely related to those which were around when life began to evolve on this planet. The relationship between pogonophoran and bacteria is symbiotic. The pogonophoran could not survive without the bacteria and the later bacteria probably could not live in the open sea without the physical protection of the pogonophoran.

Man
The gut of man is fairly heavily infested with micro-organisms. Some work has been done on the effects of these organisms, but no real benefit or harm has been established for most of them. Some people have been kept in a germ-free state for a short while and have suffered no noticeable harm. More details will be found in Chapter 10 of this book.

Termites
Another species that is totally dependent on gut micro-organisms (in this case mainly flagellate protozoans) is the wood-eating termite. In this case, both the termites and the flagellates could not exist without one another. This then is the most extreme symbiosis possible. These termites live on a diet of wood, but, like the herbivores, are unable to break down the cellulose component of the wood. The young, newly hatched termites do not have any flagellates in their gut but rapidly become infested with them by eating the faeces of adult termites. By this means, the flagellates eventually reach the hind gut of the young termites. The flagellates act in the same way as the micro-organisms in herbivores, by converting cellulose to soluble carbohydrates. In exchange for a constant supply of wood they provide the host with nourishment. Once again an anaerobic environment is provided by the termite's gut.

The flagellates have become so reliant on their host that if the termite colony should die for any reason so will the flagellates. The only members of the huge termite colony that are not dependent on the presence of flagellates in their guts are the king and queen. They are fed by the rest of the colony on a specially prepared diet. When a new colony is to be formed the king and queen fly off, carrying with them a small piece of the infested faeces so that the new colony will have a supply of flagellates.

The dependence these two species have on one another is a little more extreme than that concerning the herbivores. The queen actively makes the flagellates available to her young, whereas the cow relies on the bacteria being all around in the air.

Leguminous Plants

It is not only animals that rely upon bacteria for their nourishment; there is a family of plants, the Leguminosae, which utilise bacteria to convert nitrogen into nutrients. Nitrogen in the form of a highly stable molecule (N_2) is one of the constituents of the air, and it is washed down into the soil by the action of rain. Most plants require nitrogen for their survival and the legumes utilise the bacterium *Rhizobium* to convert this nitrogen to ammonia which can then be readily absorbed by the plant.

These bacteria are micro-organisms which normally live in the soil. If they are close to the root of a legume the relationship begins. It is thought that the plant's roots secrete a substance which activates the bacteria to change shape and enter the plant's roots via the root hairs (tiny, hair-like branches which take nutrients and water from the soil). Once inside the plant, the bacteria multiply very quickly and activate the plant root cells which divide (multiply) and form tuberous swellings called root nodules. Inside these nodules the plant and bacteria work together to fix the nitrogen to make it into a form usable by the plant.

The root nodule continues to grow and mature until eventually

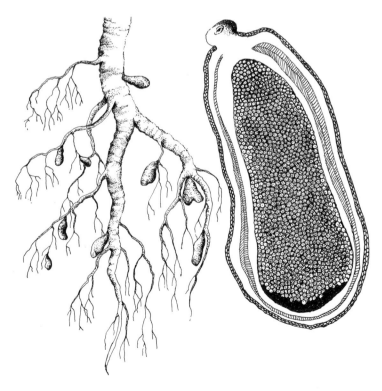

Legume root nodules. The detail on the right shows the symbiotic bacteria within a nodule.

it breaks off from the plant, releasing more bacteria into the soil. These bacteria are then ready to be taken up by another plant. The relationship is therefore very close as only the correct bacteria will work for each plant. The symbiosis between the plants and the bacteria was only discovered fairly recently although root nodules have been observed for centuries. Since the sixteenth century, farmers have noted that, by growing crops of beans, peas and clover, they can enhance the quality of their soil. This is because of the extra nitrogen fixed by the nodules. This relationship is similar in function to those of the termites and herbivores, but it is interesting to note that these phenomena occur in the plant world as well as the animal kingdom.

Fish and Squid

One more example of animals using bacteria is quite different in that bacteria in this instance act as a torch for the host and have little to do with feeding. Many different species of fish and squid emit light. Some do this from photophores (light-giving cells), but the ones that concern us here use luminescent bacteria. These bacteria live within the bodies of their host, mostly close to the head.

The bacteria live in the host's organs where they receive nourishment, a constant environment and a degree of protection from the outside world. In exchange they glow and, when grouped together, the luminescence they emit can be quite powerful. Most of the species that use the bacteria are creatures from the deep sea where light hardly penetrates or is absent altogether. The light produced by the bacteria enables the host to see where it is going and probably acts as a lure enticing small prey fishes. It is also likely that these species use the light to attract a mate.

The light from these bacteria is constant, but various species have grown 'eyelids' to enable the light to be flashed (for example the fish *Photoblepharon palpebratus*). Some fish have developed a control mechanism to regulate the output of the bacteria. The mechanism involves chromatophores, which are under direct control of the host. They can act so rapidly that the light appears to flash, thus focusing attention on the source.

In most luminescent fishes, the bacteria-filled organs lie near the eyes, but they can occur anywhere on the animal's body. The Japanese Knight Fish (*Monocentris japonicus*) has a pair of these organs on each side of the tip of its lower jaw. These presumably excite the curiosity of small fish, which on investigating fall prey to the luminescent fish. The species *Malacocephalus laevis* has light

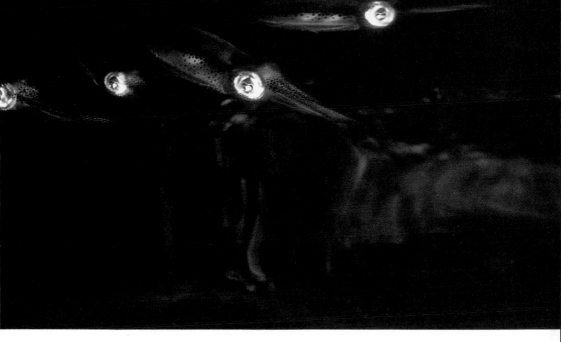

organs on its belly and argentinoid fishes have rectal light organs. Certain myopsid squids have bacterial luminescence; for example, *Sepiola loligo* has a pair of such organs in the mantle cavity, near the ink sac. Other squid have more sophisticated organs; *Sepiola birostrata* and *Loligo edulis* have developed a reflector or lens to 'magnify' the light emitted.

Most luminescent species have to be reinfected each generation as the bacteria die out with the host. The bacteria live freely in the water until they come upon a suitable host. All the species that utilise luminescent bacteria are marine; there have been no recorded incidents of freshwater species with these bacteria.

The examples mentioned above all concern plants and animals associating symbiotically with various bacteria. Some of them come upon the bacteria accidentally whereas others seek them out and make sure that their offspring will have the benefit of their use. The strength of the associations varies as well from total dependence to useful but not essential.

103

The impact of the legume symbiosis on the nitrogen cycle is considerable, and this is demonstrated when the data which records the annual amount of molecular nitrogen fixed by symbiotic associations is studied. These figures show that around 100 million tonnes of nitrogen are fixed in this way each year. Without this nitrogen the quality of the soil would be very poor, too poor to sustain the cover and variety of vegetation that exists today. The benefits to the soil of planting legumes has been known for centuries, but we still do not make good use of this knowledge to convert nitrogen into forms that can be used by all crops. The action of legumes compares well with the use of nitrogen fertilisers. A good legume crop can fix 200 kg or more of nitrogen per hectare in one year, which can be compared to the application of two tonnes of fertiliser on the same area of land. The use of plant rotation also has none of the disadvantages that the continual use of fertilisers has on the general ecology.

As can readily be seen from the contents of this chapter the actions of bacteria in conjunction with plants and animals are of fundamental importance to us all. The very existence of herbivores and soil enrichment from legumes are both of great value. Without these our world would indeed be a different place. The other bacteriological associations may not be of such great importance but they are in their own right quite fascinating and extraordinary. They show how a species, with the aid of another, can live in an environment or situation in which it would not be able to survive well alone. In this way species are able to fill in all the small, inhospitable niches provided by nature.

8 FUNGI

This chapter deals with the many relationships between plants and animals with various fungi. One such relationship, the lichen symbiosis, has already been mentioned in Chapter 6. Some of the examples resemble the bacterial associations in Chapter 7, in that the breakdown of cellulose is the main reason for the symbiosis. Other examples in this chapter concern the germination of seeds and the nourishment of forest trees.

Trees

The first example is of tremendous importance to all living things. A good deal of the world's oxygen is produced during photosynthesis by the huge forested areas of the earth. Most forest trees form nuts as their seed, which are distributed to some extent by animals; a great many of them, however, fall close to the parent tree. They then grow up in close proximity to the trees all around, gradually forming a forest. In this situation there is obviously intense competition for root space. Many trees have enormous root systems to draw enough nutrients and water from the soil to allow them to grow and mature. In a dense forest, however, this is often impossible, as there are too many trees chasing too little soil and space. It is in this situation that the trees' association with various fungi is of paramount importance. The fungi cover the tips of the tree roots with hyphal tissue, forming a cobweb-like net which penetrates the soil, effectively increasing the tree's root system and enhancing root efficiency. The effect on the tree is as though its root area has been dramatically increased, which enables more nutrient and water to be drawn from the soil. The fungi can produce much thinner strands and make a much more dense 'root' system than can the tree.

It is thought that in the initial stages of the partnership between tree and fungi, when the latter is invading the root system, the

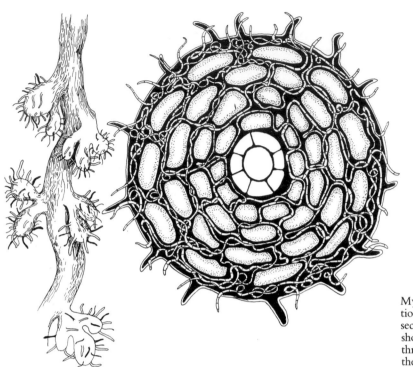

Mycorrhizal associations. The cross section on the right shows the fungal threads penetrating the root system.

fungi parasitise the tree by syphoning off nutrients from the host to enable them to grow. When the full fungal mycorrhiza is formed, however, the fungi and tree are in equilibrium and the relationship changes from parasitic to symbiotic. (This theory, however, has yet to be authenticated.) The fungi benefit from the safety of being anchored to a strong tree and share in the sugars produced by the tree roots, and the tree's root system is effectively extended. Some trees are able to form symbiotic partnerships with several different species of fungi, for example over forty species of fungus are known to associate with *Pinus sylvestris*. Several different fungi can inhabit a single tree at the same time and thus work in a form of co-operation with one another.

Without the aid of these fungi, most of our enormous, dense forests would not be able to exist. Instead the trees would grow much more sparsely and would therefore not produce anything like the quantity of oxygen that they do now. The more these vast forests are destroyed by the incursions of man the more this efficient partnership will be important. Huge forests in South America (where most of the world's oxygen is produced) are being cleared every year in the name of progress, and many scientists are seriously worried about the effects of this on global oxygen levels.

Various experiments have been conducted in order to quantify the benefits to the tree of this fungal relationship, and it has been established that the benefit is highly significant to the health and growth of the tree. It is possible therefore that it is only due to this relationship, which enabled the optimum amount of oxygen to be produced from a given area of soil, that the oxygen levels necessary to sustain life on this planet are maintained.

These fungi do not exclusively associate with trees; they can also be found in the roots of many perennial plants, annuals, ferns and liverworts.

Orchids

During the nineteenth century orchids became much prized in Europe as exotic and rare houseplants. They were seen as status symbols, along with the number of servants and the variety of transport used by a household. Orchids consequently fetched considerable prices, and thus adventurous men were encouraged to risk life and limb in tropical rain forests to collect specimens. To avoid the costly business of collecting from the wild, botanists attempted to grow the plants from seed.

Nobody had any success in germinating the tiny orchid seeds until 1904, when the French botanist Nöel Bernard showed that if the seeds were infected with various fungi they could be persuaded to germinate and grow. These were fungi that could provide the seeds with an environment enriched with carbohydrates and vitamins which because of their size they do not naturally possess. Most seeds carry enough nutrient around them to provide a suitable growing medium for the seeds until the seedlings are large enough to fend for themselves. This did not happen with the orchid seeds; they were so small that no nutrient was carried with them. The fungi do not cease affecting the plant after germination. When the seedling has developed a root system the fungi form a fine web of fungal filaments which extend the plant's roots and help it to survive until it develops shoots and leaves which allow it to photosynthesise and survive unaided. Some of the 25,000 known species of orchid have become so dependent on their fungal ally that they develop no root system of their own (Corallorhiza) and rely totally on the fungal 'roots' for their nourishment and water supply.

The orchid obviously benefits from the fungal infection both at

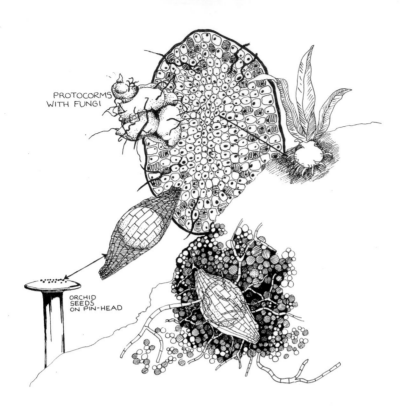

ORCHID
SEEDS
ON PIN-HEAD

The growth and
development of
orchid seeds with their
symbiotic fungi.

germination and to some extent for the rest of its life from the
extension of its root system. The fungus benefits from a stable
'home' and shares in the surplus sugars produced by the plant.

The discovery of this method of germination made orchid
plants more readily and cheaply accessible to the European grower
and encouraged hybridisation and cultivation of these extraordinary plants. Many of the species available today would not exist
unless the grower was able to cross plants with each other and raise
the seeds that are formed. Methods of germination have since been
developed which do not require fungi, but rather use modern
technology to establish germination and maintain the plants.
There are many other plants that associate with fungi in similar
ways, but it is the need for fungi at the germination stage that sets
orchids apart.

Insects

The Ambrosia Beetle (*Xyleborus ferrugineus*) uses fungi in a way
similar to the insects described in Chapter 7. The ambrosia beetle
has a long, narrow body resembling an elongated cylinder. It

inhabits most parts of the world, but is especially common in warm climates. Both adults and larvae burrow in wood, making the tunnels and galleries in which they live. As with the termites, the ambrosia beetle and its larvae are unable to digest the wood cellulose without the aid of a fungus, which they cultivate in the galleries and tunnels. The fungus cannot break down the cellulose from intact wood, but needs the beetle to chew it up into small pieces. The beetle transfers fungi from tree to tree when necessary by filling a sack on its thorax with fungal spores and 'planting' them in tunnels made in the new wood. This it does quite deliberately and so indicates some kind of awareness of the benefits of cultivating the fungi.

Both parties obviously benefit from this association. Neither can utilise the wood on which they live without the assistance of the other; thus it is a true partnership. There are other benefits, however. The beetle eats some of the fungi to keep it under control, as otherwise it might smother the galleries. It has been demonstrated recently that the ambrosia beetle is unable to pupate without the ergosterol which is synthesised by the fungus. This situation is analogous to the larvae of the large blue butterfly needing ant grubs as part of their diet before pupation can take place. The beetle is therefore much more highly dependent on the fungus than would at first appear. The fungus is provided with an even climate and protected from the outside world by the burrows in which it lives, and it is also 'gardened' by the beetle to keep it healthy. It is given a supply of chewed-up wood as nourishment.

There are several other species of beetle that cultivate fungi in similar ways, and this example has been cited as fairly typical. Not all these other beetles are so dependent on the host, as they are able to pupate without the fungi. Some beetles use fungi that are harmful to trees, and this is the reason for dutch elm disease being transmitted from tree to tree.

The next example is rather similar to that of the ambrosia beetle but sufficiently different to be worthy of mention. Termites of the family Macrotermitines associate with the fungus *Termitomyces* in order that the latter may help with the breakdown of cellulose for the former. Termites belong to the order Isoptera, which means equal (*iso*) winged (*ptera*), the fore and hind wings of the adult being the same size. These insects are often called 'white ants' because of their colour and social habits, but in fact they are not ants at all. Termites of Asia and tropical Africa use fungi to assist in the digestion of the wood which forms the bulk of their diet.

The termites, of which there several hundred species, take the chewed-off lumps of wood down into their nest and there form a cellulose rich environment on which the fungus is grown. The action of the fungus is to break down the cellulose into an easily consumable carbohydrate soup, which the termites can digest. The termites husband the fungus by keeping enough wood always ready and by eating the old and not so vigorous fungi in order to keep the colony at optimum efficiency. The fungi are thus kept fed and housed by the termites and provide digestible food in return.

Colonies of termites live in nests which are complex systems of galleries and pathways, made of soil and vegetation cemented together by saliva. Some species' nests are subterranean and some are arboreal. In the arboreal nests enormous 'hills' are made several metres high, which have become a feature of the landscape. A few days after tropical rainstorms break a long drought, there appear toadstools in the form of 'fairy rings' around the edges of these termite nests. These are the external manifestations of the fungi gardens kept by the termites. The toadstools are all that can be seen of the fungi that live in the nest chambers below ground.

Toadstools around a termite hill. The cut-away section shows how the toadstools appear from the fungal gardens within the hill.

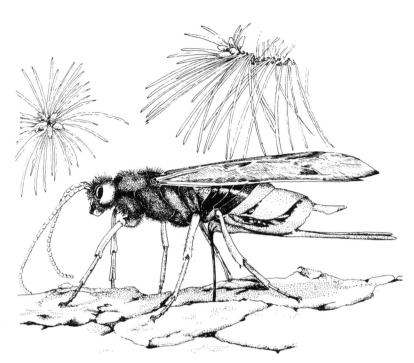

A female woodwasp using the ovipositor to lay her eggs under the bark of a conifer.

As well as eating from their fungus cellulose soup the termites provide a very important service to the rest of the world. They consume and thereby degrade vegetable litter on the ground to simple carbon compounds. This removes unwanted and rotting matter and cleans up the environment for all other animals and plants. Without this service, provided by termites and other micro-organisms and insects, the dead leaves and so on would build up to intolerable levels.

A similar relationship has developed between woodwasps of the families Lymexylidae and Siricidae, among others, with fungi. The adult wasps do not need the fungus, but the female infects her larvae with fungus when she lays her eggs. She pushes her long ovipositor into the wood where the larvae develop and at the same time pushes the fungus into the crevice. Here it breaks down the wood in the usual way, providing the larvae with food while they grow and mature.

9 ODDITIES

This chapter deals with two examples of symbiotic relationships that do not easily fall into any of the previous categories. These associations are 'odd' in the sense that they do not conform to the normal cleaner/feeder roles. They are either complex, involving a number of different species, or transitory in that the animals only associate occasionally.

Plants, Ants and Butterflies

The first example is a highly complex one, involving four species (two plants and two animals) native to South America. The Passion Flower Vine, Wild Cucumber, ants and the butterfly *Heliconius* all associate to form a symbiotic relationship.

Heliconius lives on the pollen produced by the wild cucumber plant, which is in itself unusual, as most butterflies live on nectar rather than pollen. The heliconius larvae do not like cucumber leaves as food and therefore the adult female butterfly lays her eggs on a more favoured food species, the passion flower vine. This plant attempts to protect itself from the voracious appetites of the developing larvae by encouraging ants to live on its foliage. The ants have a vicious bite and present a considerable threat to the larvae. In order that the ants may be encouraged to live on the plant, the vine has nectaries on its stem and on the undersides of its leaves. These nectaries produce a sugary fluid which attracts the 'sweet tooth' of the ants. Ants are consequently always swarming over the plant, and when the female butterfly comes to lay her eggs she is chased away. The butterfly has to avoid the main part of the flower where the ants patrol and lays her eggs on the farthest extremities of the plant. The tips of the leaves and the ends of the tendrils with which the plant clings to its support are the safest places, furthest away from the marauding ants.

The eggs hatch relatively free from danger and then the larvae begin to eat along the leaves, or tendrils, towards the main plant. As they eat they grow, and so the closer they get to the ants the bigger they are. Thus by the time they came into contact with the ants they should be large enough to defend themselves. In this way enough of them develop to form chrysalises and thereby ensure that there will be another generation of butterflies.

This is an unusual relationship in many ways. Firstly, most butterflies lay eggs on or near the plants on which the adults feed. In this instance, another plant, unrelated to the adults' food provider, is used. Secondly, it is rare for a plant to enlist the services of an animal to protect it, but the plant produces nectar for no other purpose than to entice the ants. It is quite obviously not accidental that the ants are present in such numbers. It is assumed that the passion flower vine must have suffered a great deal from the appetites of the butterfly larvae many years ago. This must have been the evolutionary incentive that caused the plant to enlist the help of the ants to protect it. Although this protection is far from complete, as many larvae survive the ants, it is better than nothing. Once such a relationship has been established it will be possible for it to continue to evolve and become more effective as a means of protection for the vine (*see* p. 114).

In this association there are two organisms in symbiosis (the passion flower vine and the ants), and two that are parasitised (the cucumber by the adult butterfly, and the passion flower vine by the butterfly larvae). The passion flower vine benefits from a degree of protection from the ants. It is much better for the vine that the larvae should eat the tips of the plant, and effectively do a bit of harmless pruning, rather than that they should munch away at the central part of the vine and possibly cause permanent damage. The ants gain a host with constant supplies of nectar and the occasional larva to eat.

Birds and Badgers

The second 'oddity' in this chapter is the relationship between a bird, the Black Throated Honey Guide (*Indicator indicator*), and the Honey Badger or Ratel (*Mellivora capensis*). It is a loose association which is not vital to either species, but very common.

The ratel lives mainly in forested areas of South East Asia and throughout most of Africa. It is omnivorous, with a varied diet of vegetation, small animals and insects. Like many other animals, the ratel is fond of sweet things, and it is this fondness that brings it

into an association with the honey guide. The bird will direct the ratel (or any other willing partner) to bees' nests and the honey they contain. The honey guide is unable to break open the nest on its own and needs a stronger animal to do this for it. The bird looks for a ratel, then hops about, calling to gain attention. It then flies to the honey and indicates where it is located. The ratel is a stocky, badger-like creature with strong fore-feet and powerful jaws. The ratel breaks open the nest and consumes the honey. This is a perfect relationship as the honey guide is not interested in eating the honey but the bee-grubs and the beeswax, both of which are left un-touched by the ratel. Although the relationship is so useful and simple the creatures do not form any lasting bond. The meetings are largely accidental, as the bird can only direct the ratel for fairly short distances. It tends to find the honey first and then look for a passing ratel. Although most frequently associated with the ratel, the honey guide does not exclusively guide this animal; it is quite able to take any willing species to the honey. African people often use the bird for the same reason, and the bird is quite willing to guide them too.

Passion flower vine with heliconius larvae and ants.

A honey guide indicating the whereabouts of a bees' nest.

The honey guide has a parasitic egg-laying habit. It lays its eggs in other birds' nests, being careful to remove or destroy the host's eggs. This method of reproduction means that the honey guides are given no parental instruction on how to guide the ratel, nor are they fed at any time on wax or bee-grubs. This guiding behaviour therefore must be instinctive.

10 MAN

The subject of symbiotic relationships between plants and animals and man is difficult. Most of the relationships for consideration are distant rather than close and often it is not essential for either species to associate. Such associations would not be considered symbiotic by those who adhere to the strictest definition of the word, but usually do enrich the lives of both parties to some extent, and are therefore mentioned here.

For convenience the associations will be split into three basic groups, although some will be seen to overlap within these group-ings. The first and most scientifically acceptable group are those micro-organisms that flourish inside and on the surface of our bodies. The second group are the relationships between man and species that are essential to him but somewhat physically removed, or the by-products of symbiotic relationships that are essential to human life—a kind of once-removed partnership. The third group are those organisms that are aesthetically or socially de-sirable to human beings and that are maintained or encouraged for our pleasure.

Micro-organisms

Humans do not require populations of bacteria to digest their food; in that way they differ from the herbivores already men-tioned. We eat relatively little cellulose and therefore its break-down is not essential. This is one of the benefits of an omnivorous diet; when a considerable variety of foods is ingested it does not matter a great deal if absolute efficiency of digestion does not occur for all types of food. We do, however, have large numbers of assorted bacteria within the alimentary canal. Most of these seem at least benign and probably beneficial; they prefer the conditions of the human (and other mammalian) gut as a place to live. It is not in their interests to harm the host, as usually if the host gets very ill or dies the micro-organisms perish too. There are some bacteria that do harm the host and cause a variety of sicknesses, but this is not the normal condition. The offending organisms are usually ingested with food or water. The constant populations of bacteria can in some cases be beneficial to man. Our endobacteria are known to be able to synthesise vitamins in excess of their own metabolic needs and it is most likely that we absorb the surplus. Adult humans probably get most of their pyridoxine and vitamin B12 in this way. They may well receive other nutritional benefits not yet established from the micro-organisms that live within their gut. As with all medical events to do with man, it is difficult to do any experiments to prove the effects of such organisms. A reverse proof of their benefit lies in the fact that animals kept 'germ-free' often require vitamin supplements to continue developing and living. It is therefore probable that this deficiency is, at least in part, as a result of the non-availability of bacterial vitamins. There is also evidence that several generations of chickens, rats and mice can be kept totally germ-free, so bacterial vitamins may not be essential to all animals. Germ-free animals have also been found to be more vulnerable to infection by pathogenic (disease-inducing) bacteria. This could be explained by the immune reaction effect, the principle on which we are injected with a miniscule amount of the bacteria to which we require immunity. This injection activates the body's immune responses and gives us protection against a later invasion of this same bacteria. The germ-free animals have no past experience of these pathogens and therefore have no natural resistance to them.

Germ-free animals' hearts seem, from experimental evidence, to be less efficient than control populations. The heart's output is measured by the volume of blood it pushes around the body. Germ-free animals' hearts move less blood than normal animals. The reason for this is not understood, but it is likely to have something to do with one or more of the micro-organisms living within the animal's (and probably man's) body.

Human beings cannot be kept in germ-free conditions artificially for medical and social reasons, so this animal data will not be confirmed from human research. Very occasionally sick humans are kept in sterile conditions, but usually only for a short while, so no lasting effects of micro-organism deprivation can be seen. As these people, by definition, are unhealthy, they are also atypical and not very useful for such research. It is most probable, from the research already complete, that we actually require several of the multitude of different micro-organisms that live within the gut for a normal healthy life. Proving a need for one particular organism is exceptionally difficult. It would require isolation of the bacterium, fungus or virus under study and the understanding of its nature and exact function and relationship with its host. We are a long way away from that situation at the moment. It is also quite possible that the effects of these tiny creatures is not that simple; it could be as a result of the action of them all working together that the host receives benefit. The effects of the organisms probably varies according to their relative numbers, the numbers and variety of other organisms present and even the body temperature and general health of the host. The increasing use of drugs, especially antibiotics, must have a considerable effect on the populations within the gut.

As can be readily seen from this information, the problems of studying intestinal organisms are legion. Specific pathogens have been studied for many years, as the incentive is great to understand how they live and the results of their actions on the host in order that the diseases they cause may be eliminated. The incentive to study benign or helpful micro-organisms is not so great. An attitude of 'let sleeping dogs lie' tends to prevail among researchers who have more pressing problems to study.

It is not only gut organisms that infest man. Large areas of our skin are covered in vast populations of fungi and bacteria. Most of these are harmless or even beneficial, but some are potential pathogens. These are harmless on the surface of the skin, but they can turn nasty if they penetrate it and invade the host's body. The same problems of study involve these organisms, but to a lesser extent. At least they can easily be removed locally and studied.

It cannot be stated categorically which of these micro-organisms on our skin or in our gut are living symbiotically with us. We obviously provide them with a suitable habitat or they would not live with us. In some cases perhaps we do more than provide a suitable environment for the individual organism, but that has yet to be established.

Indirect Relationships

Humans, along with most creatures and plants, require oxygen for their survival. As has been discussed in a previous chapter the production of oxygen, largely from forestation, enables the world to 'breathe'. By managing these forests, replanting new trees and controlling disease, man is able to assist the trees to produce oxygen. Thus we can be in a type of once-removed symbiosis both with the trees and the mycorrhizal fungi, which also help them to grow. It is highly debatable whether we are at the moment helping to produce oxygen in this way. Vast areas of forest all over the world, but particularly in South America, are cleared every year. The world's demand for wood and products made from wood increases yearly, and little account is taken by many industrialists of the global effects of their actions. It is not only the doom-watchers that are alarmed by the casual way millions of acres of forest are destroyed every year.

It could be said that humans form a kind of symbiosis with the crops and stock that they use for food. If you consider the effect on the whole species rather than the individuals that go to make it up, this statement can be seen to be true. It is only as a result of the intervention of man that many species of plant and livestock exist today. For example, several food crops, such as corn (which is no longer to be found in its wild state), would probably have become extinct if they had not been kept going by man. The same applies to camels used as beasts of burden, which no longer exist in their natural, wild state. It is not only crops that are artificially main-tained by humans, many plants kept for the aesthetic pleasure they give are nurtured in this way. In these days of a general interest in conservation, exceptionally rare plants are guarded by man from specimen gatherers and other dangers. Seeds can be gathered and grown in nurseries to prevent the species from dying out. Can this be classed as symbiosis? Humans attempt to maintain the life of a species or individual in exchange for food or beauty. It certainly could not be described as a close relationship; but in the sense that our lives would, to some extent, be impoverished by their absence I believe that such associations can be included within the widest definition of symbiosis.

The same applies to the animals we use for fur, meat, milk and eggs. From the individual lamb's point of view no symbiosis exists. The animal is not allowed to develop to maturity or reproduce, it is therefore receiving minimal benefit from the close association with man. However, from an overall species' view the association is beneficial; much greater numbers of animals are maintained than could occur naturally. Endangered species are also protected from becoming extinct. Dairy cows and egg-laying hens are slightly more obviously symbiotic with man. The cows and hens are cared for and kept healthy in exchange for the milk and eggs they produce. They benefit from the attention they receive throughout their lives until they are no longer able to produce. At the end of their useful lives they are killed, usually in a relatively humane way, which avoids any lingering or violent death such as might be suffered in the wild. Other complicating factors can creep into this argument if animal battery farms are considered. Surely if cruelty in any form is perpetrated on these animals, the symbiotic element disappears. If we do not provide a suitable and comfortable environment for our livestock, they cannot be said to benefit from their relationship with us.

Man also benefits indirectly from the soil enrichment brought about by legume root nodules. Better soil produces more abundant crops for man to harvest. Although the process of nitrogen-fixing has been understood for quite some while, it is still little used by farmers. Instead of costly and often dangerous fertilisers, crop rotation involving legumes and much less fertiliser could enrich the soil for future plantings. Thus the sowing of legumes encourages the growth of the bacteria that produce the nodules.

Aesthetics

Beauty has always been important to man, and he has generally found the world about him to be attractive. Nature has, by most cultures, been singled out as the most constant source of aesthetic pleasure. More works of art are created about the world around us than about any other single subject. As already mentioned many species of plants are kept simply for the pleasure they give. This is also true of some animals; for example, doves and peafowl are kept just so that their owners can observe them. They are not really pets as there is little or no physical contact between the animals and their owners. Different cultures keep different animals for their beauty; the goldfish has been kept for centuries in the east, and whole new strains have been developed by man to accentuate the 'good' points and minimise the less favoured features.

The keeping of pets has also an interesting history. The reasons why humans associate so closely with these animals has long been discussed. It is generally believed that the animals are kept for companionship, to show affection and to help educate children. It is also arguable whether the animals benefit much from their relationship with man. They are often housed, fed and fussed over in the most inappropriate way. This can obviously cause the animal suffering and in these cases the relationship is not symbiotic. If however, the animal is cared for properly and allowed to live a comfortable and reasonable life, it benefits from the association.

There is a fine line between these two conditions, and much of it is open to argument. Does a dog really benefit from a ten-mile walk every day? Animals usually expend only the amount of energy necessary to catch prey, find water and so on; they rarely do more. It is difficult to state that some pets are symbiotic with man as opinions on how they should be cared for will differ from person to person. It seems likely that symbiosis is clear in some of these relationships. Once more the argument is complicated by man's interference with species in the form of 'line' breeding and the development of new strains of animals. Is it justifiable to breed a variety of dog that has restricted nasal passages just because the breeder likes to see dogs with pushed-in noses? Is it ethical to breed a dog with so much surplus skin that it wrinkles all around its face and often causes skin problems? These and many more breeds that are 'made' by man for his own pleasure and often suffer in consequence show that it is not through altruism that humans keep pets. Aesthetic considerations are placed above common sense and the suffering of our pets.

There are many animals that have learned to live close to man and take whatever benefit they can from the association. Davenport says in his contribution to *Symbiosis* by Henry, 'It is quite clear that in many symbiotic species specific behaviour has evolved that is different from that of their free living relatives.' This is true of many of the species discussed in previous chapters and also of such animals that share closely human dwellings and habitats. Birds learn to make nests in thatched roofs, foxes scavenge in city centres, bats live in belfries and bridges, storks use chimneys, barn owls use barns—the list is endless. All these species have learned to tolerate humans, and furthermore they use man and his dwellings, food and so on to improve their own quality of life. When humans encourage these associations a degree of symbiosis exists. Barn Owls (*Tyto alba*) used to be cave-dwellers, but now live in barns because of the readily available food source there. Farmers encourage them because they help to control rodent populations. For the same reason farmers often keep cats, and in the United States the Rat Snake (*Elaphe obsoleta*) is encouraged to live in farm outbuildings so that it will eat rats and mice.

Humans get a great deal of pleasure from the beauty of their environment, but have strange ways of showing it. Most people would agree that coral is beautiful, but apparently they have no compunction about tipping effluent and other pollutants, which cause the coral to die, into the sea. The extraordinary variety of plants and animals also give pleasure, but man is still content to allow hundreds of species to become extinct each year. It would be lovely to say that man is in some kind of symbiosis, however vague, with all the flora and fauna of this planet, but he is demonstrably not so. There are many plants and animals that are considered pests, but invariably they are an essential food source for others that are deemed beautiful; they are therefore important when put into context. When enlightenment is more universal about the importance of plants and animals perhaps man will be truly symbiotic with all those he comes into contact with, or at least benign towards them.

BIBLIOGRAPHY

Books

Abbey, E., *Cactus Country*, Time Life, New York 1977

Bates, R., *Intimate Strangers* (transcript of a BBC *Horizon* programme), BBC Publications 1976

Buchsbaum, R., *Animals Without Backbones* 1 & 2, Penguin 1964

Burton, J., *Animals of the African Year*, Eurobook Ltd 1972

Caullery, M., *Parasitism and Symbiosis*, Sidgwick and Jackson, London 1952

Friedrich, H., *Marine Biology*, Sidgwick and Jackson, London 1969

Friedrich, H., *Man and Animal*, Paladin 1972

Gotto, R.V., *Marine Animals—Partnerships and other Associations*, English University Press Ltd, London 1969

Hancross, D., *Master Builders of the Animal World*, Hugh Evelyn, London 1973

Harvey, E.N., *Bioluminescence*, Academic Press, New York 1952

Henry, S.M., *Symbiosis* Vols. 1 & 2, Academic Press, New York 1966

Herring, P.J., *Bioluminescence in Action*, Academic Press, New York 1978

Huxley, J., *Evolution, the Modern Synthesis*, George Allen and Unwin, London 1974

Keeble, F., *Plant-Animals*, Cambridge University Press 1910

Margulis, L., *Symbiosis in Cell Evolution*, W.H. Freeman & Co., San Francisco 1981

Nicol, J.A.C., *The Biology of Marine Animals*, Pitman, London 1960

Paturi, F., *Nature, Mother of Invention*, Thames and Hudson, London 1976

Read, C.P., *Parasitology and Symbiology: An Introduction*, Wiley, New York 1970

Reinheimer, H., *Symbiosis: A Socio-physiological Study of Evolution*, Headly Bros., London 1920

Rickelefs, R.E., *Ecology*, Thomas Nelson and Sons, London 1973

Rosebury, T., *Life on Man*, Secker and Warburg, London 1969

Rosebury, T., *Microorganisms Indigenous to Man*, McGraw-Hill, London 1962

Scott, G.D., *Plant Symbiosis*, Edward Arnold Ltd, USA, 1969

Smith, D.C., *Symbiosis of Algae with Invertebrates*, Oxford Biology Readers, Oxford 1973

Smith, D.C., *The Lichen Symbiosis*, Oxford Biology Readers, Oxford 1973

Smith, D.C., *Symbiosis in the Microbial World*, Wiley 1978

Street, P., *Animal Partners and Parasites*, David and Charles, London 1975

Thomas, L., *Medusa and the Snail*, Allen Lane, London 1980

Younge, P.J., *Bioluminescence in Action*, Academic Press, New York 1957

Articles

Arber, W., 1963, 'Bacteriophage Lysogeny', *Symp. Soc. Gen. Microbiol. (Camb.)* 13

Davenport, D., 'Specificity and Behaviour in Symbiosis', *Q. Rev. Biol.* 30:29–46

Droop, M.R., 1963, 'Symbiotic Associations', *Symp. Soc. Gen. Microbiol. (Camb.)* 13

Dubos, R., 1976, 'Symbiology between Earth and Humankind', *Science (Wash.)* 193:459

Hanks, J.H., 1966, 'Host Dependent Microbes', *Bacteriol. Rev.* 30:114–35

Horowitz, N.H. and Miller, S.L., 1962, 'Current Theories on the Origin of Life', *Fortschr. Chem. Org. Naturst.* 20:423–459

Hungate, R.E., 1955, 'Mutualistic Intestinal Protozoa' *in* Hunter and Woof (eds) *Biology and Physiology of Protozoa*, Academic Press, New York

Jennings, D.H. and Lee, D.L., 1975, *Symp. Soc. Exp. Biol.* No. XXIX

Koch, A., 'Intracellular Symbiosis in Insects', *Ann. Rev. Microbiol.* 14:21

Lederberg, J., 1952, 'Cell Genetics and Heredity Symbiosis', *Physiol. Rev.* 32:403

Limbaugh, C., 1961, 'Cleaning Symbiosis', *Sci. Am.* 205:42–49

Margulis, L., 1971, 'Symbiosis and Evolution', *Sci. Am.* 225(2):48–57

124

Nutman and Mosse, 1963, 'Symbiotic Associations', *Symp. Soc. Gen. Microbiol. (Camb.)* 13

Shaffer, 1965, 'Studies in Fatal Hypoglycemia in Axenic Piglets', *Proc. Soc. Exp. Biol. Med.* 118 : 566–570

Smith, D.C., 1979, 'From Extracellular to Intracellular, the Extent of Symbiosis', *Trans. R. Soc.* London B204 : 115–130

Smith, D.C., 1981, 'The Symbiotic Way of Life', *Brit. Mycol. Soc.*

Timourian, H., 'Symbiotic Emergence of Metazoans', *Nature, Lond.* 226 : 283–284

INDEX